Contents

T0348897

Acknowledgments

Writing a book is a large task and requires support from numerous people, and those people deserve thanks. First, I thank LeeAnn, my devoted wife of more than 20 years. She has been an unflagging fan, a counselor, and a demanding editor. She taught me much of what I have managed to learn about how to express a thought in ink. Thanks to my mother who was sure I would grow into someone in whom she would be proud when facts should have dissuaded her. Thanks also to my father for his insistence that I obtain a college education; that privilege was denied to him, an intelligent man born into a family of modest means.

I am grateful for the education provided by Virginia Tech. *Go Hokies*. The basics of electrical engineering imparted to me over my years at school allowed me to grasp the concepts I apply regularly today. I am grateful to Mr. Emory Pace, a tough professor who led me through numerous calculus courses and, in doing so, gave me the confidence on which I would rely throughout my college career and beyond. I am especially grateful to Dr. Charles Nunnally; having arrived at university from a successful career in industry, he provided my earliest exposure to the practical application of the material I strove to learn. I also thank Dr. Robert Lorenz of the University of Wisconsin at Madison, who introduced me to observers some years ago. His instruction has been enlightening and practical. Several of his university courses are available in video format and are recommended for those who would like to extend their knowledge of controls. In particular, readers should consider ME 746, which presents observers and numerous other subjects.

I thank those who reviewed the manuscript for this book. Special thanks goes to Dan Carlson for his contributions to almost every chapter contained herein. Thanks also to Eric Berg for his numerous insights. Thanks to the people of Kollmorgen Corporation (now, Danaher Corporation), my long-time employer, for their continuing support in writing this book. Finally, thanks to Academic Press, especially to Joel Claypool, my editor, for the opportunity to write this edition and for editing, printing, distributing, and performing the myriad other tasks required to publish a book.

Safety

This book discusses the operation, commissioning, and troubleshooting of control systems. Operation of industrial controllers can produce hazards such as the generation of

- large amounts of heat,
- high voltage potentials,
- movement of objects or mechanisms that can cause harm,
- the flow of harmful chemicals,
- flames, and
- explosions or implosions.

Unsafe operation makes it more likely for accidents to occur. Accidents can cause personal injury to you, your co-workers, and other people. Accidents can also damage or destroy equipment. By operating control systems safely, you decrease the likelihood that an accident will occur. *Always operate control systems safely!*

You can enhance the safety of control-system operation by taking the following steps:

1. Allow only people trained in safety-related work practices and lock-out/tag-out procedures to install, commission, or perform maintenance on control systems.
2. Always follow manufacturer recommended procedures.
3. Always follow national, state, local, and professional safety code regulations.
4. Always follow the safety guidelines instituted at the plant where the equipment will be operated.

5. Always use appropriate safety equipment. Examples of safety equipment are protective eyewear, hearing protection, safety shoes, and other protective clothing.
6. Never override safety devices such as limit switches, emergency stop switches, light curtains, or physical barriers.
7. Always keep clear from machines or processes in operation.

Remember that any change of system parameters (for example, tuning gains or observer parameters), components, wiring, or any other function of the control system may cause unexpected results such as system instability or uncontrolled system excitation.

Remember that controllers and other control-system components are subject to failure. For example, a microprocessor in a controller may experience catastrophic failure at any time. Leads to or within feedback devices may open or short closed at any time. Failure of a controller or any control-system component may cause unanticipated results such as system instability or uncontrolled system excitation.

The use of observers within control systems poses certain risks including that the observer may become unstable or may otherwise fail to observe signals to an accuracy necessary for the control system to behave properly. Ensure that, on control-system equipment that implements an observer, the observer behaves properly in all operating conditions; if any operating condition results in improper behavior of the observer, ensure that the failure does not produce a safety hazard.

If you have any questions concerning the safe operation of equipment, contact the equipment manufacturer, plant safety personnel, or local governmental officials such as the Occupational Health and Safety Administration.

Always operate control systems safely!

Chapter 1

Control Systems and the Role of Observers

I n this chapter . . .

- Introduction to observer operation and benefits
- Summary of this book

1.1 Overview

Control systems are used to regulate an enormous variety of machines, products, and processes. They control quantities such as motion, temperature, heat flow, fluid flow, fluid pressure, tension, voltage, and current. Most concepts in control theory are based on having sensors to measure the quantity under control. In fact, control theory is often taught assuming the availability of near-perfect feedback signals. Unfortunately, such an assumption is often invalid. Physical sensors have shortcomings that can degrade a control system.

There are at least four common problems caused by sensors. First, sensors are expensive. Sensor cost can substantially raise the total cost of a control system. In many cases, the sensors and their associated cabling are among the most expensive components in the system. Second, sensors and their associated wiring reduce the reliability of control systems. Third, some signals are impractical to measure. The objects being measured may be inaccessible for such reasons as harsh environments and relative motion between the controller and the sensor (for example, when trying to measure the temperature of a motor rotor). Fourth, sensors usually induce significant errors such as stochastic noise, cyclical errors, and limited responsiveness.

Observers can be used to augment or replace sensors in a control system. Observers are algorithms that combine sensed signals with other knowledge of the control system to produce *observed* signals. These observed signals can be more accurate, less expensive to produce, and more reliable than sensed signals. Observers offer designers an inviting alternative to adding new sensors or upgrading existing ones.

This book is written as a guide for the selection and installation of observers in control systems. It will discuss practical aspects of observers such as how to tune an observer and what conditions make a system likely to benefit from their use. Of course, observers have practical shortcomings, many of which will be discussed here as well. Many books on observers give little weight to practical aspects of their use. Books on the subject often focus on mathematics to prove concepts that are rarely helpful to the working engineer. Here the author has minimized the mathematics while concentrating on intuitive approaches.

The author assumes that the typical reader is familiar with the use of traditional control systems, either from practical experience or from formal training. The nature of observers recommends that users be familiar with traditional (nonobserver-based) control systems in order to better recognize the benefits and shortcomings of observers. Observers offer important advantages: they can remove sensors, which reduces cost and improves reliability, and improve the quality of signals that come from the sensors, allowing performance enhancement. However, observers have disadvantages: they can be complicated to implement and they expend computational resources. Also, because observers form software control loops, they can become unstable under certain conditions. A person familiar with the application of control systems will be in a better position to evaluate where and how to use an observer.

The issues addressed in this book fall into two broad categories: design and implementation. Design issues are those issues related to the selection of observer techniques for a given product. How much will the observer improve performance? How much cost will it add? What are the limitations of observers? These issues will help the control-systems engineer in deciding whether an observer will be useful and in estimating the required resources. On the other hand, implementation issues are those issues related to the installation of observers. Examples include how to tune an observer and how to recognize the effects of changing system parameters on observer performance.

1.2 Preview of Observers

Observers work by combining knowledge of the plant, the power converter output, and the feedback device to extract a feedback signal that is superior to that which can be obtained by using a feedback device alone. An example from everyday life is when an experienced driver brings a car to a rapid stop. The driver combines knowledge of the applied stopping power (primarily measured through inertial forces acting on the driver's body) with prior knowledge of the car's dynamic behavior during braking. An experienced driver knows how a car should react to braking force and uses that information to bring a car to a rapid but controlled stop.

The principle of an observer is that by combining a measured feedback signal with knowledge of the control-system components (primarily the plant and feedback system), the behavior of the plant can be known with greater precision than by using the feedback signal alone. As shown in Figure 1-1, the observer augments the sensor output and provides a feedback signal to the control laws.

In some cases, the observer can be used to enhance system performance. It can be more accurate than sensors or can reduce the phase lag inherent in the sensor. Observers can also provide observed disturbance signals, which can be used to improve disturbance response. In other cases, observers can reduce system cost by augmenting the performance of a low-cost sensor so that the two together can provide performance equivalent to a higher cost sensor. In the extreme case, observers can eliminate a sensor altogether, reducing sensor cost and the associated wiring. For example, in a method called acceleration feedback, which will be discussed in Chapter 8, acceleration is observed using a position sensor and thus eliminating the need for a separate acceleration sensor.

Observer technology is not a panacea. Observers add complexity to the system and require computational resources. They may be less robust than physical sensors, especially when plant parameters change substantially during operation. Still, an observer applied with skill can bring substantial performance benefits and do so, in many cases, while reducing cost or increasing reliability.

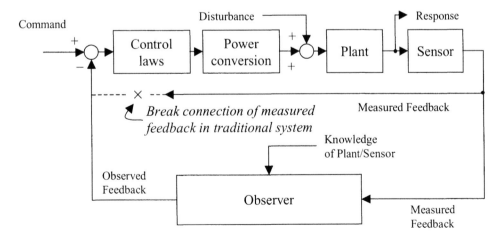

Figure 1-1. Role of an observer in a control system.

1.3 Summary of the Book

This book is organized assuming that the reader has some familiarity with controls but understanding that working engineers and designers often benefit from review of the basics before taking up a new topic. Thus, the next two chapters will review control systems. Chapter 2 discusses practical aspects of control systems, seeking to build a common vocabulary and purpose between author and reader. Chapter 3 reviews the frequency domain and its application to control systems. The techniques here are discussed in detail assuming the reader has encountered them in the past but may not have practiced them recently.

Chapter 4 introduces the Luenberger observer structure, which will be the focus of this book. This chapter will build up the structure relying on an intuitive approach to the workings and benefits of observers. The chapter will demonstrate the key advantages of observers using numerous software experiments.

Chapters 5, 6, and 7 will discuss the behavior of observer-based systems in the presence of three common nonideal conditions. Chapter 5 deals with the effects of imperfect knowledge of model parameters. Chapter 6 deals with the effects of disturbances on observer-based systems, and Chapter 7 discusses the effects of noise, especially sensor noise, on observer-based systems.

Chapter 8 discusses the application of observer techniques to motion-control systems. Motion-control systems are unique among control systems, and the standard Luenberger observer is normally modified for those applications. The details of the necessary changes, and several applications, will be discussed.

Throughout this book, software experiments are used to demonstrate key points. A simulation environment, *Visual ModelQ*, developed by the author to aid those studying control systems, will be relied upon. More than two dozen models have been developed to demonstrate key points and all versions of *Visual ModelQ* can run them. Visit www.qxdesign.com to download a limited-capability version free of charge; detailed instructions on setting up and using *Visual ModelQ* are given in Chapter 2.

Readers wishing to contact the author are invited to do so. Write gellis@qxdesign.com or visit the Web site www.qxdesign.com. Your comments are most welcome. Also, visit www.qxdesign.com to review errata, which will be regularly updated by the author.

Chapter 2

Control-System Background

I n this chapter . . .

- Common control-system structures
- Eight goals of control systems and implications of observer-based methods
- Instructions for downloading *Visual ModelQ*, a simulation environment that is used throughout this book
- Introductory *Visual ModelQ* software experiments

2.1 Control-System Structures

The basic control loop includes four elements: a control law, a power converter, a plant, and a feedback sensor. Figure 2-1 shows the typical interconnection of these functions. The command is compared to the feedback signal to generate an error signal. This error signal is fed into a control law such as a proportional-integral (PI) control to generate an excitation command. The excitation command is processed by a power converter to produce an excitation. The excitation is corrupted by a disturbance and then fed to a plant. The plant response is measured by a sensor, which generates the feedback signal.

There are numerous variations on the control loop of Figure 2-1. For example, the control-law is sometimes divided in two with some portion placed in the feedback path. In addition, the command path may be filtered. The command path may be differentiated and added directly (that is, without passing through the control laws) to the excitation command in a technique known as feed-forward. Still, the diagram of Figure 2-1 is broadly used and will be considered the basic control loop in this book.

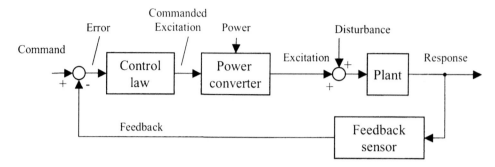

Figure 2-1. Basic control loop.

2.1.1 Control Laws

Control laws are algorithms that determine the desired excitation based on the error signal. Typically, control laws have two or three terms: one scaling the present value of the error (the proportional term), another scaling the integral of the error (the integral term), and a third scaling the derivative of the error (the derivative term). In most cases a proportional term is used; an integral term is added to drive the average value of the error to zero. That combination is called a PI controller and is shown in Figure 2-2.

When the derivative or D-term is added, the PI controller becomes PID. Derivatives are added to stabilize the control loop at higher frequencies. This allows the value of the proportional term to be increased, improving the responsiveness of the control loop. Unfortunately, the process of differentiation is inherently noisy. The use of the D-term usually requires low-noise feedback signals and low-pass filtering to be effective. Filtering reduces noise but also adds phase lag, which reduces the ultimate effectiveness of the D-term. A compromise must be reached between stabilizing the loop, which requires the phase advance of differentiation, and noise attenuation, which retards phase. Usually such a compromise is application specific. Note that

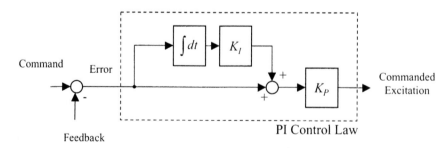

Figure 2-2. PI control law.

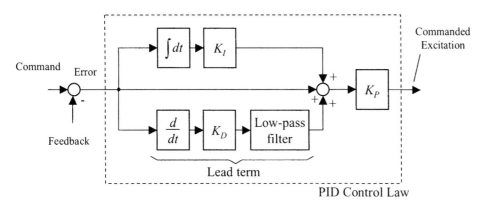

Figure 2-3. PID control law.

when a derivative term is placed in series with a low-pass filter, it is sometimes referred to as a *lead* network. A typical PID controller is shown in Figure 2-3.

Other terms may be included in the control law. For example, a term scaling the second derivative can be used to provide more phase advance; this is equivalent to two lead terms in series. Such a structure is not often used because of the noise that it generates. In other cases, a second integral is added to drive the integral of the error to zero. Again, this structure is rarely used in industrial controls. First, few applications require driving the integral of error to zero; second, the additional integral term makes the loop more difficult to stabilize.

Filters are commonly used within control laws. The most common purpose is to reduce noise. Filters may be placed in line with the feedback device or the control-law output. Both positions provide similar benefits (reducing noise output) and similar problems (adding phase lag and thus destabilizing the loop). As discussed above, low-pass filters can be used to reduce noise in the differentiation process. Filters can be used on the command signal, sometimes to reduce noise and other times to improve step response. The improvement in step response comes about because, by removing high-frequency components from the command input, overshoot in the response can be reduced. Command filters do not destabilize a control system because they are outside the loop. A typical PI control law is shown in Figure 2-4 with three common filters.

While low-pass filters are the most common variety in control systems, other filter types are used. Notch filters are sometimes employed to attenuate a narrow band of frequencies. They may be used in the feedback or control-law filters to help stabilize the control loop in the presence of a resonant frequency, or they may be used to remove a narrow band of unwanted frequency content from the command. Also, phase-advancing filters are sometimes employed to help stabilize the control loop similar to the filtered derivative path in the PID controller.

Control laws can be based on numerous technologies. Digital control is common and is implemented by programmable logic controllers (PLCs), personal computers

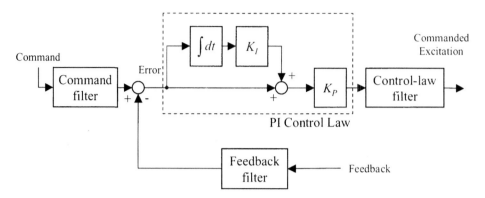

Figure 2-4. PI control law with several filters in place.

(PCs), and other computer-based controllers. Because the flexibility of digital controllers is almost required for observer implementation and because the control law and observer are typically implemented in the same device, examples in this book will assume control laws are implemented digitally.

2.1.2 Power Conversion

Power conversion is the process of delivering power to the plant as called for by the control laws. Four common categories of power conversion are chemical heat, electric voltage, evaporation/condensation, and fluid pressure. Note that all these methods can be actuated electronically and so are compatible with electronic control laws.

Electronically or electrically controlled voltage can be used as the power source for power supplies, current controllers for motors, and heating. For systems with high dynamic rates, power transistors can be used to apply voltage. For systems with low dynamic rates, relays can be used to switch power on and off. A simple example of such a system is an electric water heater.

Pressure-based flow-control power converters often use valves to vary pressure applied to a fluid-flow system. Chemical power conversion uses chemical energy such as combustible fuel to heat a plant. A simple example of such a system is a natural-gas water heater.

2.1.3 Plant

The plant is the final object under control. Most plants fall into one of six major categories: motion, navigation, fluid flow, heat flow, power supplies, and chemical processes. Most plants have at least one stage of integration. That is, the input to the plant is integrated at least once to produce the system response. For example, the temperature of an object is controlled by adding or taking away heat; that heat is

TABLE 2-1 TRANSFER FUNCTIONS OF TYPICAL PLANT ELEMENTS

Electrical

Voltage (E) and current (I)

Inductance (L)	$E(s) = Ls \times I(s)$	$e(t) = L \times di(t)/dt$
Capacitance (C)	$E(s) = 1/C \times I(s)/s$	$e(t) = e_0 + 1/C \int i(t)dt$
Resistance (R)	$E(s) = R \times I(s)$	$e(t) = R \times i(t)$

Translational mechanics

Position (P), Velocity (V), and Force (F)

Spring (K)	$V(s) = s/K \times F(s)$ or	$v(t) = 1/K \times df(t)/dt$ or
	$P(s) = 1/K \times F(s)$	$p(t) = p_0 + 1/K \times f(t)$
Mass (M)	$V(s) = 1/M \times F(s)/s$	$v(t) = v_0 + 1/M \int f(t)dt$
Damper (c)	$V(s) = F(s)/c$	$v(t) = f(t)/c$

Rotational mechanics

Rotary position (θ), Rotary velocity (ω), and Torque (T)

Spring (K)	$\omega(s) = s/K \times T(s)$ or	$\omega(t) = 1/K \times dT(t)/dt$ or
	$\theta(s) = 1/K \times T(s)$	$\theta(t) = \theta_0 + 1/K \times T(t)$
Inertia (J)	$\omega(s) = 1/J \times T(s)/s$	$\omega(t) = w_0 + 1/J \int T(t)dt$
Damper (b)	$\omega(s) = T(s)/b$	$\omega(t) = T(t)/b$

Fluid mechanics

Pressure (P) and fluid flow (Q)

Inertia (I)	$P(s) = sI \times Q(s)$	$p(t) = I \times dq(t)/dt$
Capacitance (C)	$P(s) = 1/C \times Q(s)/s$	$p(t) = p_0 + 1/C \int q(t)dt$
Resistance (R)	$P(s) = R \times Q(s)$	$p(t) = R \times q(t)$

Heat flow

Temperature difference (J) and heat flow (Q)

Capacitance (C)	$J(s) = 1/C \times Q(s)/s$	$j(t) = j_0 + 1/C \int q(t)dt$
Resistance (R)	$J(s) = R \times Q(s)$	$j(t) = R \times q(t)$

integrated through the thermal mass of the object to produce the object's temperature. Table 2-1 shows the relationships in a variety of ideal plants.

The pattern of force, impedance, and flow is repeated for many physical elements. In Table 2-1, the close parallels between the categories of linear and rotational force, fluid mechanics, and heat flow are evident. In each case, a forcing function (voltage, force, torque, pressure, or temperature difference) applied to an impedance produces a flow (current, velocity, fluid flow, or thermal flow). The impedance takes three forms: resistance to the integral of flow (capacitance or mass), resistance to the derivative of flow (spring or inductance), and resistance to the flow rate (resistance or damping).

Table 2-1 reveals a central concept of controls. Controllers for these elements apply a *force* to control a *flow*. When the flow must be controlled with accuracy, a feedback sensor is often added to measure the flow; control laws are required to combine the feedback and command signals to generate the force. This results in the structure shown in Figure 2-1; it is this structure that sets control systems apart from other disciplines of engineering.

2.1.4 Feedback Sensors

Feedback sensors provide the control system with measurements of physical quantities necessary to close control loops. The most common sensors are for motion states (position, velocity, acceleration, and mechanical strain), temperature states (temperature and heat flow), fluid states (pressure, flow, and level), and electromagnetic states (voltage, current, charge, light, and magnetic flux). The performance of most traditional (nonobserver) control systems depends, in large part, on the quality of the sensor. Control-system engineers often go to great effort to specify sensors that will provide responsive, accurate, and low-noise feedback signals. While the plant and power converter may include substantial imperfections (for example, distortion and noise), such characteristics are difficult to tolerate in feedback devices.

2.1.4.1 Errors in Feedback Sensors

Feedback sensors measure signals imperfectly. The three most common imperfections, as shown in Figure 2-5, are intrinsic filtering, noise, and cyclical error.

The intrinsic filtering of a sensor limits how quickly the feedback signal can follow the signal being measured. The most common effect of this type is low-pass filtering. For all sensors there is some frequency above which the sensor cannot fully respond. This may be caused by the physical structure of the sensor. For example, many thermal sensors have thermal mass; time is required for the object under measurements

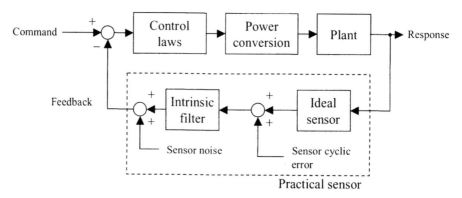

Figure 2-5. A practical sensor is a combination of an ideal sensor and error sources.

to warm and cool the sensor's thermal mass. Filtering may also be explicit as in the case of electrical sensors where passive filters are connected to the sensor output to attenuate noise.

Whatever the source of the filtering, its primary effect on the control system is to add phase lag to the control loop. Phase lag reduces the stability margin of the control loop and makes the loop more difficult to stabilize. The result is often that system gains must be reduced to maintain stability in order to accommodate slow sensors. Reducing gains is usually undesirable because both command and disturbance response degrade.

Cyclical error is the repeatable error that is induced by sensor imperfections. For example, a strain gauge measures strain by monitoring the change in electrical parameters of the gauge material that is seen when the material is deformed. The behavior of these parameters for ideal materials is well known. However, there are slight differences between an ideal strain gauge and any sample. Those differences result in small, repeatable errors in measuring strain. Since cyclical errors are deterministic, they can be compensated out in a process where individual samples of sensors are characterized against a highly accurate sensor. However, in any practical sensor some cyclical error will remain. Because control systems are designed to follow the feedback signal as well as possible, in many cases the cyclical error will affect the control-system response.

Stochastic or nondeterministic errors are those errors that cannot be predicted. The most common example of stochastic error is high-frequency noise. High-frequency noise can be generated by electronic amplification of low-level signals and by conducted or transmitted electrical noise commonly known as electromagnetic interference (EMI). High-frequency noise in sensors can be attenuated by the use of electrical filters; however, such filters restrict the response rate of the sensor as discussed above. Designers usually work hard to minimize the presence of electrical noise, but as with cyclical error, some noise will always remain. Filtering is usually a practical cure for such noise; it can have minimal negative effect on the control system if the frequency content is high enough so that the filter affects only frequency ranges well above where phase lag is a concern in the application.

The end effect of sensor error on the control system depends on the error type. Limited responsiveness commonly introduces phase lag in the control system, reducing margins of stability. Noise makes the system unnecessarily active and may reduce the perceived value of the system or keep the system from meeting a specification. Deterministic errors corrupt the system output. Because control systems are designed to follow the feedback signal (including its deterministic errors) as well as possible, deterministic errors will carry through, at least in part, to the control-system response.

2.1.5 Disturbances

Disturbances are undesired inputs to the control system. Common examples include load torque in a motion-control system, changes in ambient temperature for a temperature controller, and 50/60-Hz noise in a power supply. In each case, the primary concern is that the control law generate plant excitation to reject (i.e., prevent response to) these inputs. A correctly placed integrator will totally reject direct-current (DC) disturbances. High tuning gains will help the system reject alternating-current (AC) disturbance inputs, but will not reject those inputs entirely.

Disturbances can be either deterministic or stochastic. Deterministic disturbances are those disturbances that repeat when conditions are duplicated. Such disturbances are predictable. Stochastic disturbances are not predictable.

The primary way for control systems to reject disturbances is to use high gains in the control law. High gains force the control-system response to follow the command despite disturbances. Of course, there is an upper limit to gain values because high gains reduce system stability margins and, when set high enough, will cause the system to become unstable.

2.1.5.1 Measuring Disturbances

In the case where the control-system gains have been raised as high as is practical, disturbance rejection can still be improved by using a signal representing the disturbance in a technique known as disturbance decoupling [11, Chap. 7; 26; 27]. Disturbance decoupling, as shown in Figure 2-6, is a cancellation technique where a signal representing the disturbance is fed into the power converter in opposition to the effect of the disturbance. For the case of ideal disturbance measurement and ideal power

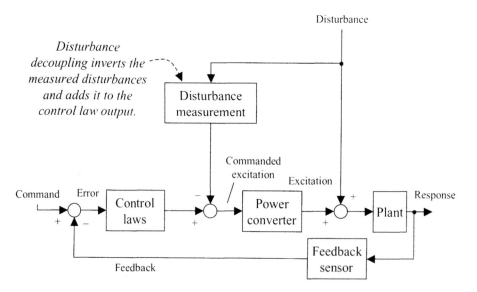

Figure 2-6. Typical use of disturbance decoupling.

conversion, disturbance decoupling eliminates the effects of the disturbance entirely. However, for practical systems, the effect of disturbance decoupling is to improve, but not eliminate, response to disturbances; this is especially true in the lower frequencies where the disturbance sensor and the power converter are often close to ideal.

For most control systems, direct measurement of disturbances is impractical. Disturbances are usually difficult to measure and physical sensors carry with them numerous disadvantages, especially increasing system cost and reducing reliability. One of the key benefits of observers is that disturbance signals can often be observed with accuracy without requiring additional sensors. For many applications, only modest computational resources must be added to implement such an observer. This topic will be discussed in detail in Chapters 6 and 8.

2.2 Goals of Control Systems

Control systems must fulfill a complicated combination of requirements. A large set of goals must be considered because no single measure can provide a satisfactory assessment. In fact, no single set of goals can be defined for general use because of the variation between applications. However, many common goals are broadly used in combination. In this section, eight common goals for control systems will be discussed. In addition, the role of observers in helping or, in some cases, hindering the realization of those goals will be discussed.

2.2.1 Competitively Priced

Control systems, like almost all products in the industrial market, must be delivered at competitive prices. The virtues of a control system will be of little value if the application can be served equally well by a less expensive alternative. This is not to say that a customer will not pay a premium for enhanced performance. However, the manufacturer offering premium products must demonstrate that the premium will improve the cost–value position of the final product.

Arguments for observer-based methods can be at either end of the cost–value spectrum. For example, if an observer is used to help replace an existing sensor with one that is less expensive, the argument may be that for a modest investment in computational resources, sensor cost can be reduced. In other cases, it can be argued that observers increase value; for example, value could be increased by providing a more reliable feedback signal or a more accurate feedback signal that will lead to improved performance.

Those readers who are leading their companies in the use of observers should expect that they will have to demonstrate the practical advantages of observers if they want the methods to be adopted. Bear in mind that observers often produce undesirable characteristics, such as increased computational costs. At the very least, they require time to develop and training for staff or customers to learn new methods.

2.2.2 High Reliability

Control systems must be reliable. A proven way to enhance reliability is by reducing component count, especially connectorized cables. Electrical contacts are among the least reliable components in many systems. Observers can increase reliability when they are used to eliminate sensors and their cables.

Observers are not the only alternative for removing sensors. There is a wide variety of techniques to remove sensors, usually by measuring ancillary states; for example, the hard-disk-drive industry long ago began employing *sensorless* technology, eliminating commutation-position[1] sensors in PC hard disks by measuring the electrical parameters of the motor driving the disk. This points out that *sensorless* is actually a misnomer; sensorless applications normally eliminate one sensor by relying on another. Still, the results are effective. In the case of sensorless hard drives, the position sensor and its cabling eliminated.

Observers offer a key enhancement for sensorless operation. The problem with most sensorless schemes is that the signals being measured usually have poor signal-to-noise characteristics, at least in some operating conditions. Returning to the example of a hard-disk controller, direct (that is, nonobserver-based) voltage measurement works well in the disk-drive industry where motor speeds are high so that the voltages created by the motor are relatively large. These same techniques work poorly at low speeds so that they cannot be used in many applications.[2] Because observers combine the sensed signals (which may have high noise content) with the model signals (which are nearly noise free), they can remove noise from the calculated output, greatly extending the range of sensorless operation. So observers can be the best alternative to allow the elimination of sensors in some applications, and thus, they can be an effective way to simultaneously increase system reliability and reduce cost.

2.2.3 Stability

Control systems should remain stable in all operating conditions. The results of unstable operation are unpredictable; certainly, it is never desirable and in many cases, people may be injured or equipment damaged. In addition to maintaining absolute stability, systems must maintain reasonable margins of stability. For example, a temperature controller with low margins of stability may respond to a commanded

---

[1] Note that this discussion relates to position sensing for commutation, the process of channeling current to produce torque in a motor. Commutation requires only coarse sensing, often just a dozen or so positions around the disk. Hard-disk drives use an additional track on the disk itself for the fine position sensing, which allows the much more accurate location of data on the disk surface.

[2] This voltage, called the back-electro-motive force or back-EMF, is produced by motors in proportion to the moving magnetic field of the motor. In most cases, the back-EMF is proportional to the speed of the motor. Thus, at low speed, the back-EMF signal is low and noise has a greater effect.

temperature change of 5° by generating oscillatory changes of 5° or 10° that die out only after minutes of ringing. Such a system may meet an abstract definition of stability, but it would be unacceptable in most industrial applications. Margins of stability must be maintained so that performance can be predictable. Two common measures of stability, phase margin and gain margin, will be discussed in Chapter 3.

Observers can improve stability by reducing the phase lag within the control loop. For example, the process of converting a sensor signal often involves filtering or other sources of phase lag. In the motion-control industry, it is common to use the simple difference of two position samples to create a velocity signal. Such a process is well known to inject a time delay of half the sample time. By using an observer this phase delay can be removed. In applications requiring the highest performance, the removal of this phase lag can be significant.

2.2.4 Rapid Command Response

Command response measures how well the response follows a rapidly changing command. Most control systems follow slowly changing commands well but struggle to follow more rapidly changing signals. In most cases, it is considered an advantage for a control system to follow rapid commands accurately.

A key measure of system response is bandwidth. The bandwidth is defined as the frequency where the small signal response falls to 70.7% of the DC response. To find the bandwidth of a control system, create a sinusoidal command at a relatively low frequency and measure the amplitude of the response. Increase the frequency until the amplitude of the response falls to 70% of the low-frequency value; this frequency is the bandwidth.

The most common way to improve command response is to raise the gains of the control laws. Higher gains help the system follow dynamic commands but simultaneously reduce margins of stability. Tuning, the process of setting control-law gains, is often a compromise between command response and margins of stability. As discussed above, observers can increase margins of stability and thus allow incrementally higher gains in the control law.

2.2.5 Disturbance Rejection

Disturbance rejection is a measure of how well a control system resists the effect of disturbances. As with command response, higher gains help the system reject disturbances, but they reduce margins of stability. Again, tuning control-law gains requires a compromise of response and stability.

Observers can help disturbance rejection in two ways. As with command response, disturbance response can be improved incrementally through higher control-law gains when the observer allows the removal of phase lag. Second, as discussed in Section 2.1.5.1, observers can be used to observe disturbances, allowing the use of disturbance decoupling where it otherwise might be impractical.

2.2.6 Minimal Noise Response

Noise response is a measure of how much the control system responds to noise inputs. The problem may be in the plant response where the concern is that the noise unduly corrupts the system output. On the other hand, the concerns may be with noise generated by the power converter. Noise fed into the control law via the command, feedback, and control-laws calculations is transferred to the power converter where it can create high-frequency perturbations in the power output. That noise can be objectionable even if the plant filters the effect so much that it does not measurably affect the system response. For example, noise in a power supply may generate high-frequency current perturbations that cause audible noise. Such noise may make the noise generation unacceptable, even if final filtering components on the power supply output remove the effect of the noise on the power supply's output voltage.

The concern with noise response is usually focused on response to high-frequency signals. High system gain is desirable at lower frequencies. A control system is expected to be responsive to signals at and below the system bandwidth. Well above the bandwidth, high gain becomes undesirable. The output does not respond to the input in any useful way (because it is greatly attenuated), but it still passes high-frequency noise, generating undesirable perturbations, audible noise, and unnecessary power dissipation. Lower gain at frequencies well above the bandwidth is equivalent to improved (reduced) noise response.

The first step to reducing noise response is reducing the amplitude of the noise feeding the control system. This may come by improving system wiring, increasing resolution of digital processes, or improving power supply quality to the sensors and control laws. After this path has been exhausted, the next step is usually to filter noise inputs. Filters are effective in reducing noise, but when filters are in the control loop, they add phase lag, reducing margins of stability; control-law gains often must be reduced to compensate. Since margins of stability must be maintained at an acceptable level, the end effect is that filtering often forces control-law gains down.

Observers can exacerbate problems with sensor-generated noise. One reason is that one of the primary benefits of observers is supporting increased control-law gains through the reduction of phase lag. The increase of control-law gains will directly increase the noise susceptibility of the typical control system. In addition, observers often amplify sensor noise above the bandwidth of the sensor. The details of this effect are complicated and will be explained in Chapter 7. For the present, readers should be aware that observers often will not work well in systems where sensor noise is a primary limitation.

2.2.7 Robustness

Robustness is a measure of how well a system maintains its performance when system parameters vary. The most common variations occur in the plant. As examples, the capacitance of a power supply storage capacitor may vary over time, the rotational inertia of a mechanism may vary during different stages of machine operation, and the amount of fluid in a fluid bath may vary and change the thermal mass of the bath. The control system must remain stable and should maintain consistent performance through these changes. One challenge of observer-based techniques is that robustness can be reduced by their use. This is because observers rely on a model; when the plant changes substantially and the model is not changed accordingly, instability can result. Thus, robustness should be a significant concern any time observers are employed.

2.2.8 Easy Setup

Control systems should be easy to set up. One of the realities of modern industry is that the end users of control systems are often unfamiliar with the principles that make those systems work. This can be hard for control-system designers to accept. It limits the use of novel control methods because those people further down the product-use chain (for example, technicians, salespeople, and end users) may not fully understand why these methods are useful or how they should be configured. Certainly, observers fit into this class of solutions. In many cases, after they have been implemented, tested, and shown to be effective, they still must be clearly explained to be ultimately successful. In addition, designers must strive to keep observers easy to set up. Observers are software-based closed loops with control laws that must be tuned; as will be discussed, this process can be simplified by careful design.

2.3 *Visual ModelQ* Simulation Environment

When learning control-system techniques, finding equipment to practice on is often difficult. As a result, designers must often rely on computer simulations. To this end, the author developed *Visual ModelQ*, a stand-alone, graphical, PC-based simulation environment, as a companion to this book. The environment provides time-domain and frequency-domain analysis of analog and digital control systems. *Visual ModelQ* is an enhancement of the original *ModelQ* in that *Visual ModelQ* allows readers to view and build models graphically. More than two dozen *Visual ModelQ* models were developed for this book. These models are used extensively in the chapters that follow. Readers can run these experiments to verify results and then modify parameters and other conditions so they can begin to experiment with observers.

Visual ModelQ is written to teach control theory. It makes convenient those activities that are necessary for studying controls. Control-law gains are easy to change. Plots of frequency-domain response (Bode plots) are run with the press of a button. The models in *Visual ModelQ* run continuously, similar to the way real-time controllers run. The simulated measurement equipment runs independently so parameters can be changed and the effects seen immediately.

2.3.1 Installation of *Visual ModelQ*

Visual ModelQ is available at www.qxdesign.com. The unregistered version is available free of charge. While the unregistered version lacks several features, it can execute all the models used in this book. Readers may elect to register their copies of *Visual ModelQ* at any time; see www.qxdesign.com for details.

Visual ModelQ runs on PCs using Windows 95, Windows 98, Windows 2000, or Windows NT. Download and run the executable file *setup.exe* for *Visual ModelQ V6.0* or later. Be aware that the original version of *ModelQ* is not compatible with *Visual ModelQ*. Note that *Visual ModelQ* comes with an online help manual. After installation, read this manual. Finally, check the Web site from time-to-time for updated software.

2.4 Software Experiments: Introduction to *Visual ModelQ*

The following section will review a few models to introduce the reader to *Visual ModelQ*.

2.4.1 Default Model

When *Visual ModelQ* is launched, the default model is automatically loaded. The purpose of this model is to provide a simple system and to demonstrate a few functions. The default model and the control portion of the *Visual ModelQ* environment are shown in Figure 2-7.

The model compilation and execution are controlled with the block of three buttons at the upper left of the screen: compile (green circle), stop execution (black

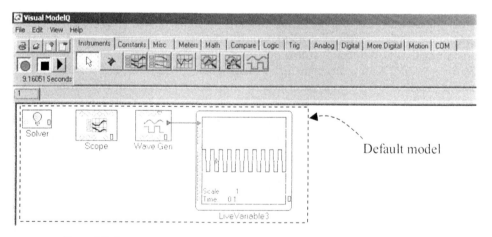

Figure 2-7. Screen capture of *Visual ModelQ* environment showing the default model.

Figure 2-8. Compile and run controls.

square), and start execution (black triangle). These blocks, with the current execution time (here, 9.16051 seconds), are shown in Figure 2-8. If a model must be compiled before it can be run, the green circle will turn red. The circle will turn red at launch and anytime either a block or a wire is added to or taken away from the model. Any time a model is recompiled, the model timer will return to 0 seconds and all default values of model blocks will be reloaded.

The default model is detailed in Figure 2-9. There are four blocks, two of which are connected with a wire:

- *Solver*: The solver configures the differential-equation solver used to simulate system components. Note: One and only one solver is required for every model.
- *Scope*: The main scope provides a display for up to eight channels of input. The workings of the scope and its trigger mechanism are similar to those of a physical oscilloscope. Note: At least one scope is required for every model.
- *Waveform Generator*: The waveform generator can be used to generate standard waveforms such as sine waves and triangle waves. Frequency, amplitude, and phase are all adjustable while the model is running. The generator here is set to produce a square wave at 10 Hz.

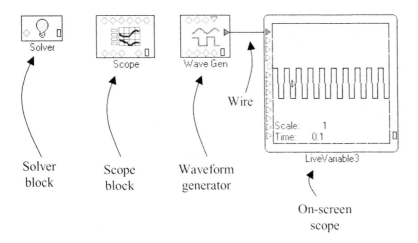

Figure 2-9. Detail of default model.

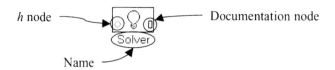

h node — Documentation node

Name

Figure 2-10. Detail of *Solver* node.

- *Live Scope*: The *Live Scope* displays its output on the block diagram. *Live Scope* variables automatically display on all main scope blocks as well. Notice in Figure 2-7 that a short wire connects the output of the waveform generator to the input of the scope; this connection specifies that the *Live Scope* should plot the output of the waveform generator.

2.4.1.1 Viewing and Modifying Node Values

Blocks have nodes, which are used to configure and wire the elements into the model. For example, the solver block, shown in Figure 2-10, has two nodes. There is a configuration node (a green diamond) at the left named *h*. This node sets the sample time of the differential-equation solver. The sample time is set to $10\,\mu s$ by default.

The solver block includes a documentation node (a rectangle) at the right. The documentation node, which is provided on almost all *Visual ModelQ* blocks, allows the user to enter notes about the block for reference. The name of the block, *Solver* in this case, is shown immediately below the block. The user can change the name of any *Visual ModelQ* block by positioning the cursor within the name and double-clicking.

There are several ways to read the values of nodes such as the *h* node of the sample block. The easiest is to use *fly-over* help. After the model is compiled, position the cursor over the node and the value will be displayed in a *fly-over* block for about a second, as shown in Figure 2-11.

The value of configuration nodes can be set in two ways. One way is to place the cursor over the node and double-click. The *Change/View* dialog box is then displayed as shown at the top right of Figure 2-12. The value can be viewed and changed from this dialog box.

The second way to set values is to use the *Block set-up* dialog box. Right-click in the body of the block; this brings up a pop-up menu as is shown center left in

h = 1E-5

Figure 2-11. *Visual ModelQ* provides fly-over help for nodes.

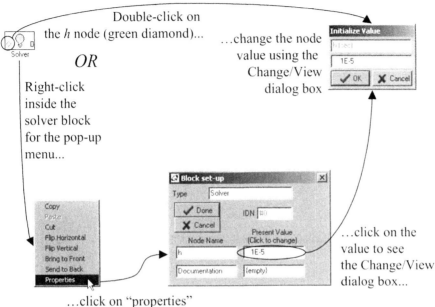

Figure 2-12. Two ways to change the h parameter of the solver block.

Figure 2-12. Select the *Properties* item in that menu to bring up the *Block set-up* dialog box. This box will show the value of all the nodes in the block. Click on the value to bring into view the *Change/View* dialog box.

2.4.1.2 The WaveGen Block

The *WaveGen* block has ten nodes, as shown in Figure 2-13. The nodes are:

- Waveform: Select initial value from several available waveforms such as sine or square waves.
- Frequency: Set initial frequency in Hertz.
- Amplitude: Set initial value of peak amplitude. For example, setting the amplitude to 1 produces an output of ±1.
- Enable: Allows automatic disabling of the waveform generator. When the value is 1, the generator is enabled. When 0, the generator is disabled. For digital inputs such as this node, *Visual ModelQ* considers any value greater than 0.5 to be equivalent to 1 (true); all values less than or equal to 0.5 are considered equivalent to 0 (false). This function will be especially useful when taking Bode plots since all waveform generators should be disabled in this case.
- Output: Output signal of waveform generator.

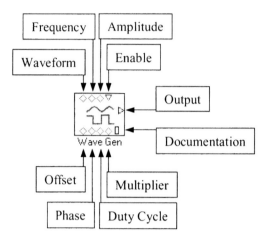

Figure 2-13. The waveform generator has 10 nodes.

- Offset: Initial value by which the waveform generator output should be offset.
- Phase: Initial value of waveform phase, in degrees, of the waveform generator. For example, if the output is a sine wave, the output will be:

$$\text{Output} = \text{Amplitude} * \sin(\text{Frequency} \times 2\pi \times t + \text{Phase} \times \pi/180) + \text{Offset}.$$

- Duty cycle: Initial value of percentage duty cycle for pulse waveforms.
- Multiplier: Value by which to multiply waveform generator output. This is normally used for unit conversion. For example, most models are coded in *Systeme International* (*SI*) units. If the user finds RPM more convenient for viewing than the SI radians/second, the multiplier can be set to 0.105 to convert RPM (the user units) to radians/second (SI units). The multiplier node is present in most instruments such as scopes and waveform generators to simplify conversion to and from user to SI units.

The *Enable* node of the *WaveGen* block is an input node, as the inward-pointing triangle indicates. Input nodes can be changed while the model is running and they can be wired in the model. Neither of these characteristics is true of configuration nodes (those shaped like diamonds).

Using the block set-up dialog box can speed the setup of more complicated blocks such as the *WaveGen*. The *WaveGen* block set-up dialog is shown in Figure 2-14. The benefit of the block set-up dialog is that all of the parameters are identified by name and can be set one after the other. Notice that the first node in the dialog, *Output*, cannot be changed (the button at right allows only "*View . . .*"). This is necessary because some nodes, such as output nodes, cannot be configured manually.

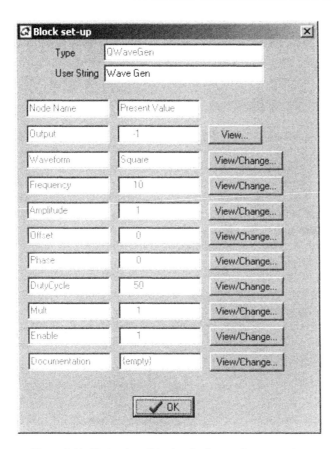

Figure 2-14. Block set-up dialog box for the waveform generator.

The parameters of the waveform generator set in the nodes are only initial (precompiled) values. To change the configuration of the waveform generator when the model is running, double-click anywhere inside the block and bring up the real-time *WaveGen* control panel. This panel, shown in Figure 2-15, allows six parameters of the waveform to be changed while the model is running. The buttons marked "<" and

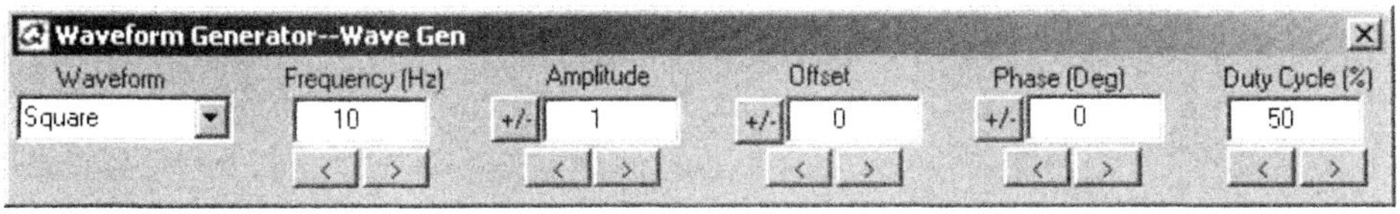

Figure 2-15. Waveform generator control panel which is displayed by double-clicking on the *WaveGen* block after the model has been compiled.

">" move the value up and down by about 20% for each click. Changing these values has no permanent affect on the model; each time the model is recompiled, these values will be returned to the initial values as specified by the nodes.

2.4.1.3 The Scope Block

The Scope block, with a list of its nodes, is shown in Figure 2-16. Most of the nodes set functions that are consistent with laboratory oscilloscopes and thus will be familiar to most readers. One node that should be discussed is the *Trigger Source* node. This node sets the initial variable that will trigger the scope when the scope mode is set to *Auto* or *Normal*. If this variable is not set prior to compiling the model, a warning will be generated. To eliminate this warning, simply double-click on the node and select a variable from a drop-down list to trigger the scope. Choose from any *Variable* or *Live Scope*, as shown in Figure 2-16.

The scope display is normally not visible. However, it can be made visible by double-clicking inside the scope block after the model has been compiled. The block can be made not visible by clicking the "X" icon at the top right of the scope window.

The scope display provides two tabs: *Scale* and *Trigger*. The *Scale* tab (shown in Figure 2-17) provides control of the horizontal and vertical scaling. The Trigger tab provides various trigger settings. At the bottom of the scope there are a few controls. Starting at the bottom left of Figure 2-17:

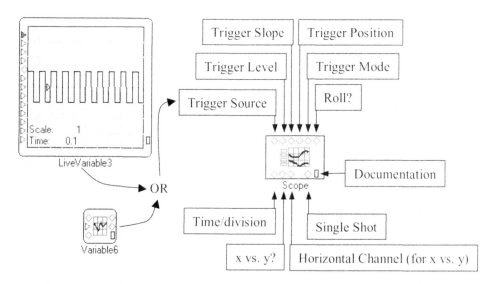

Figure 2-16. The *Trigger Source* of a *Scope* can be set to any variable (such as *Variable6*) or any *Live Variable* (such as *LiveVariable3*).

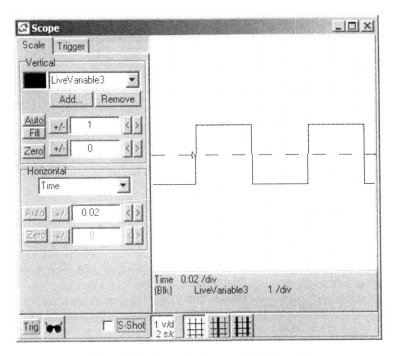

Figure 2-17. Output of main scope in default model.

- the *Trig* button flashes green for each trigger event;
- the sunglasses button hides the control panel at left, maximizing the display area of the plot;
- the single-shot check box enables single-shot mode;
- the scale-legend control button turns the scale legend (immediately below the plot) on and off;
- the three cursor buttons select 0, 1, or 2 cursors.

Note that single-shot mode stops the model from running after the scope screen has filled up. Restart the model using the *Run* (black triangle) button after each single-shot event.

2.4.1.4 *The* Live Scope *Block*

The default model also includes a *Live Scope* block, as shown in Figure 2-18. The input comes in at top left, with the scale, offset, and time scale set in the nodes just below that. The *Show* node determines whether the variable in the *Live Scope* is displayed in the main scopes after each compile (note that variables that display in a

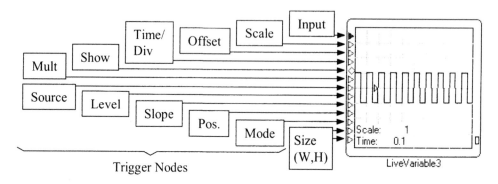

Figure 2-18. Detail of the *Live Scope* nodes.

Live Scope also can be displayed in any main scope block). The *Mult* node specifies a multiplier, which scales the variable before plotting.

The next five nodes are trigger nodes. The *Trigger Source* node specifies the signal that triggers the *Live Scope*. If this variable is unwired, the *Input* (first) node will be used as the trigger. Most of the remaining nodes have equivalent functions on standard oscilloscopes except the last two nodes, *Width* and *Height*, which set the size of the *Live Scope* block in pixels.

Live Scopes provide simple display features compared to the main scope block, and there are several limitations. No more than two channels can be displayed using a *Live Scope*. There are fewer trigger options. Another limitation is that *Live Scopes* only show input vs time; there is no option for Input1 vs Input2 (*x* vs *y*) as there is for the main *Scope* blocks.

The *Live Scope* also has several advantages. First, the wiring to a *Live Scope* makes it clear which variable is being plotted; this makes the display more intuitive, especially in larger models. Second, because the result is displayed on the model, it is often easier to convey information to others using the *Live Scope*. It is this reason that caused the author to prefer the *Live Scope* to the standard scope throughout this book. Finally, almost all of the *Live Scope* parameters are input nodes, and all input nodes can be wired into the circuit. This means that a model can be constructed to automatically change those values as the model executes.

2.4.2 Experiment 2A: Simple Control System

The remainder of this chapter will discuss three experiments written to introduce the reader to control-system modeling in *Visual ModelQ*. Experiment 2A is a simple control system. The model diagram is shown in Figure 2-19. The model is comprised of several elements:

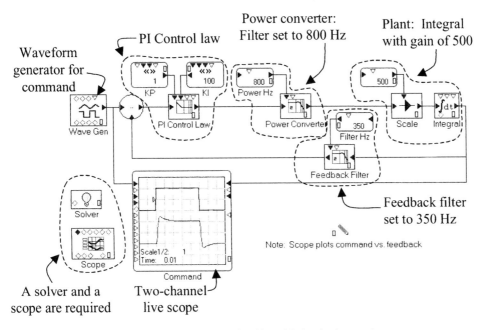

Figure 2-19. Experiment 2A: *Visual ModelQ* model of a simple control system.

- A waveform generator, which produces the command.
- A summing junction, which compares the command and the feedback (output from the feedback filter) and produces an error signal.
- A PI control law, which is configured with two *Live Constants*, a proportional gain, K_P, and an integral gain, K_I. These blocks will be discussed shortly.
- A filter simulating the power converter. The power converter is a two-pole low-pass filter set for a bandwidth of 800 Hz and with a zeta (damping ratio) of 0.707.
- An integrating plant with an intrinsic gain of 500.
- A filter simulating the feedback conversion process. The feedback filter is a two-pole low-pass filter set with a bandwidth of 350 Hz and with a zeta of 0.707.
- A two-channel *Live Scope* that plots command (above) against actual plant output (below).
- A solver and scope, both of which are required for a valid *Visual ModelQ* model.

2.4.2.1 Visual ModelQ *Constants: Many Ways to Change Parameters*

Visual ModelQ provides numerous ways to change model parameters. Of course, any unwired node can be changed by double-clicking on a node or right-clicking and bringing up the *Block set-up* dialog box (see Figure 2-12). However, numerous blocks are provided to simplify the task of changing node values.

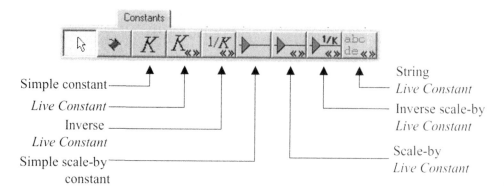

Figure 2-20. Selecting from among the many constants available in *Visual ModelQ*.

The Constants tab in the *Visual ModelQ* environment (top of Figure 2-7) currently provides seven constant types: simple constants, standard and inverse *Live Constants*, simple scaling constants, standard and inverse scaling *Live Constants*, and string constants. The selection buttons for each of these constants are shown in Figure 2-20, which is a screen capture of the top portion of the *Visual ModelQ* environment.

Live Constants, such as K_P and K_I in the PI controller of Figure 2-19, provide the most control. The icons of blocks have a "≪ ≫" symbol. After the model has compiled, double-click anywhere inside the block and the adjustment box of Figure 2-21 will appear. Using the adjustment box, the value of the parameter can be changed while the model runs. A new value can be typed in with the keyboard by clicking the cursor in the value edit box. (Note that when using the keyboard, the new value does not take effect until the enter key is hit.) In addition, there are six logarithmic adjustment buttons in the adjustment box. The double less-than block (≪) reduces the value to the next lowest value with the first digit being 1, 2, or 5. For example, if the value of the variable is 1.75, clicking "≪" will change the value to 1, clicking again will reduce it to 0.5, clicking again will reduce it to 0.2, and so on. Each click reduces the value approximately by half. The double greater-than (≫) performs a similar function except it moves to the next higher value: 1, 2, 5, 10, 20, and so on.

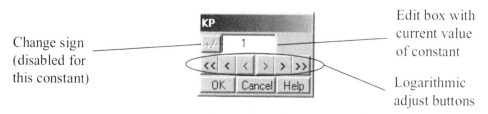

Figure 2-21. Adjustment box appears when double-clicking a *Live Constant* any time after the model has been compiled.

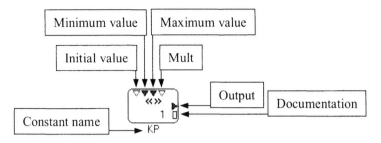

Figure 2-22. Detail of a *Live Constant*.

The remaining adjust buttons are straightforward. The bold single less-than button reduces the value of the variable by about 20% for each click; the nonbold single less-than button reduces the value by about 4%. The bold and nonbold single greater-than blocks perform a similar function, only raising the value. If the parameter can take on values of both signs, the +/– button will be enabled, allowing a change in sign at the click of a button.

The *Live Constant* model block and its nodes are shown in Figure 2-22. The initial value node specifies the value that the constant is reset to after each compile. The minimum and maximum nodes specify the range that the input can take on. The multiplier and documentation nodes are standard *Visual ModelQ* nodes. The output makes available the value of the *Live Constant* so it can be wired in the model. The value displayed in text inside the block is not scaled by the *Mult* node, while the value in the output node is.

2.4.2.2 Inverse Live Constants

The inverse *Live Constant* works like the standard *Live Constant* except the output is one divided by the parameter value and then multiplied by the value of the *Mult* node (Figure 2-23). This constant is used when the model needs to scale by the inverse ($1/x$) of the parameter value such as is usually the case for mass, moment of inertia, thermal mass, capacitance, inductance, and many other physical parameters. The inverse *Live Constant* is a space-saving alternative to combining a standard *Live Constant* and a $1/x$ block.

1/K indicates "Inverse *Live Constant*."

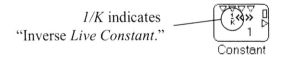

Figure 2-23. Inverse *Live Constant* generates divided output.

(a)   (b)

Figure 2-24. Standard (a) and inverse (b) scale-by *Live Constants*.

2.4.2.3 *Scale-by* Live Constants

Scale-by *Live Constants* are similar to *Live Constants* except that the output node is the product of an input node and the value of the constant. In fact, if the input node is set to one, scale-by *Live Constants* behave identically to standard *Live Constants*. The two scale-by *Live Constants* (standard and inverting) are shown in Figure 2-24.

2.4.2.4 *String Constants*

String constants allow the model constants to be adjusted as strings. The user selects a string from a list and the string *Live Constant* block outputs an integer value according to the position in the list occupied by the selected string. For example, the string *Live Constant* named *Select XN* in Figure 2-25 is configured to allow the user to select one of four strings: *X0*, *X1*, *X2*, and *X3*. Depending on which constant the user selects, the output node will be 0, 1, 2, or 3, according to the position of the string within the list.

The string *Live Constant* node is configured with two input nodes on the left side of the block. The *Strings* node should be filled first; double-click on this node and type in a list of string constants. The *Strings* node dialog box for the block of Figure 2-25 is shown in Figure 2-26. There is no specific limit on constant length or on the number of strings that one string *Live Constant* can hold. Next, select the *Live Constant's* initial string by double-clicking on the upper left node.

The string *Live Constant* is often used with an analog switch, as is the case in Figure 2-25. The switch has a control node at top center, the value of which determines which of the four inputs at left is routed to the output: moving from top to bottom, 0 selects the first input, 1 the second, and so on. In the case of Figure 2-25, *Select XN* is equal to *X1*, as is indicated inside the string *Live Constant* block. This produces an output of 1, which is fed to the switch control node. That causes the Position-1 (second) input node to be connected to the output node of the switch block. Figure 2-25 also has an *Inspector* block, which can display the value of any node. Here, the inspector shows that output of *Switch4* is 10.01, matching the value of the *Live Constant X1* which is connected to the Position-1 input node. Note that the naming of the four *Live Constants* at left to match the list of strings in *Select XN* is for clarity and has no effect on the operation of the model.

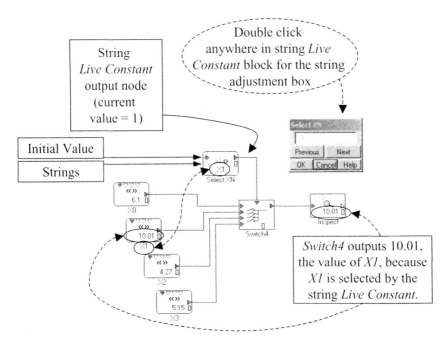

Figure 2-25. A string *Live Constant* outputs an integer based on user-selected character strings.

2.4.2.5 Simple Constants

The last *Visual ModelQ* constants are the simple constants. These constants are similar to the *Live Constant*. However, the simple constants do not support the adjustment box of Figure 2-21; changes to the value are made via double-clicking on the input

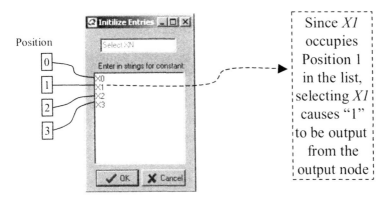

Figure 2-26. The user types in a string list to configure the string *Live Constant* block.

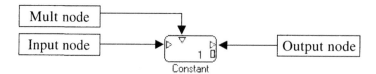

Figure 2-27. The simple constant.

node. (Note that changing a node value is permanent after the model is saved.) Also, neither maximum nor minimum limits can be set. The simple constants take a little less screen space than a *Live Constant* in the model diagram. Use the simple constant for parameters that are changed only occasionally. The simple constant is shown in Figure 2-27; the simple scale-by constant adds scaling.

2.4.2.6 Hot Connections on a Live Scope

The *Live Scope* supports a feature called *hot connection*. Anytime the model is running, double-click on the *Live Scope* and a Live Scope control panel will appear. Click "Hot Connect" to close the dialog box; move the mouse over any wire or input–output node in the model and click. The Live Scope will temporarily graph the value of that node or wire; the scope outline will turn green to indicate that the scope is in *hot connect* mode. (For two-channel *Live Scopes*, Channel 1 displays the hot connection; Channel 2 is unaffected.) Click "Restore Scope" in the Live Scope control panel or recompile the model to restore the scope to its original display. Note that the scope scale and offset nodes may need to be adjusted to view the signal; any changes to scaling and offset will be restored when the scope is restored. The operation of hot connection is displayed in Figure 2-28. Hot connections are especially useful when debugging a model, as wires and nodes can be viewed without adding a scope, which forces recompilation.

2.4.3 Command Response and Control-Law Gains

Visual ModelQ is designed to simplify the process of evaluating the effects of parameter value variation. This is a common need when modeling control systems, for example, in the tuning process. Tuning is the adjustment of control-loop gains to achieve optimal performance. It is often carried out in working systems (and in models) by observing the effect of numerous incremental changes of control-law gains. For example, in Experiment 2A, K_P might be adjusted up and down in small steps while observing the effect on the step response. Experiment 2A is constructed to make this process fast and simple.

The two components of Experiment 2A that simplify tuning are the *Live Constant* and the *Live Scope*. After model compilation, double-clicking on the *Live Constant* named K_P brings up the K_P adjustment box, which allows rapid changes of value,

The *Live Scope* normally shows *Command,*
but *Hot Connect* can temporarily display power converter output,
without recompiling or even stopping the model.

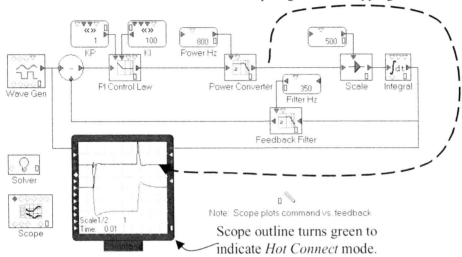

Note: Scope plots command vs. feedback

Scope outline turns green to
indicate *Hot Connect* mode.

Figure 2-28. Hot connect allows temporary reconfiguration of *Live Scopes* while the model executes.

perhaps one per second. Compare this to standard modeling environments where the model must be stopped, modified, and recompiled. A simple change can take on the order of a minute. In addition, the *Live Scope* gives immediate feedback of the effect of the new parameter, without the need for the user to issue a command to display a plot. To experiment, launch *Visual ModelQ*. Click *File, Open . . .* to open the model *Experiment_2A.mqd*. Click *Run*. Double-click on the K_P block and use the $\ll$ and $\gg$ buttons to move the value up and down. The results should be equivalent to those shown in Figure 2-29.

2.4.4 Frequency Domain Analysis of a Control System

Control systems often need to be analyzed in the frequency domain. The most intuitive method of frequency-domain analysis for most people is the Bode plot, which graphs gain and phase across a range of frequencies. A gain plot displays the amplitude of an output signal divided by the amplitude of the input signal at many frequencies as if sine waves at many frequencies had been applied to the model. A phase plot displays the time lag of the output compared to the input for many sine waves. In the laboratory, the instrument that is commonly used to generate Bode plots is called a dynamic signal analyzer (DSA). *Visual ModelQ* provides a DSA, which is used regularly in Chapters 4 through 8. Experiment 2B, shown in Figure 2-30, is

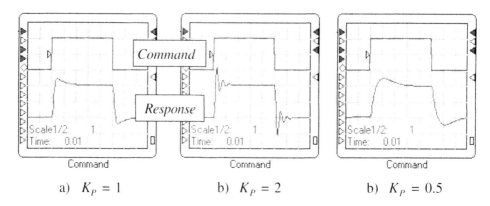

a) $K_P = 1$ b) $K_P = 2$ b) $K_P = 0.5$

Figure 2-29. Results of varying K_P in Experiment 2A.

Experiment 2A modified to include a DSA, which is shown just right of the wave-form generator.

2.4.4.1 *The* Visual ModelQ *DSA*

The DSA is wired in line with the excitation path. In most cases, the DSA is used to analyze command response and so will normally be inserted in line with the command as it is in Figure 2-30. All DSAs read all model variables, no matter how they are wired. In *Visual ModelQ*, the term *variables* includes three types of signals:

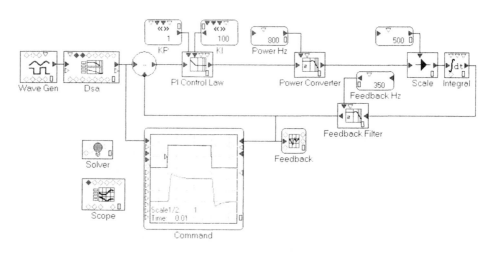

Figure 2-30. Experiment 2B: Experiment 2A with a DSA.

- The input to 1-channel *Live Scopes*,
- The input to Channel 1 of 2-channel *Live Scopes*, such as *Command* in Figure 2-30, and
- *ModelQ* variables blocks such as *Feedback* in Figure 2-30.

The DSA here will be used to show the relationship between command and feedback. Notice that Experiment 2B required the addition of the variable block *Feedback* at top right. In Experiment 2A that node was not connected to a variable block as it was only needed for display as Channel 2 of a *Live Scope*. In Experiment 2B, an explicit variable block named *Feedback* is required to grant access of the signal to the DSA.

2.4.4.2 DSA Nodes

The complete details on configuring a DSA go beyond the scope of this chapter. However, a few details should be mentioned to prepare the reader for the use of DSAs in this book. The four most important nodes of a DSA block are shown in Figure 2-31. At left is the input node. Normally, the DSA is inactive and the input node passes directly to the output node. However, when the user wants a new Bode plot, the DSA is commanded to excite the model. This temporarily disconnects the input node and replaces it with a random signal excitation. The *Excitation Amplitude* and *DSA Inactive* nodes will be discussed in Section 2.4.4.5.

2.4.4.3 The DSA Display

A Bode plot from a DSA is shown in Figure 2-32. This shows the relationship between command and feedback, commonly called the *closed-loop* response, for Experiment 2B where $K_P=1$ and $K_P=2$; the gain plots are above and the phase plots below. Most of the time, the closed-loop gain plot will be of primary interest. The two cases here behave similarly at low frequency (shown at left) and the plots below about 100 Hz are nearly indistinguishable. However, above 100 Hz, there are significant differences, especially where the gain of the $K_P=2$ case sharply rises before falling, displaying an undesirable characteristic called *peaking*. The purpose of this section is to introduce *Visual ModelQ*, so a detailed discussion of resulting waveforms is outside the scope

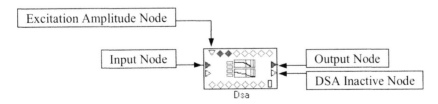

Figure 2-31. Detail of DSA nodes.

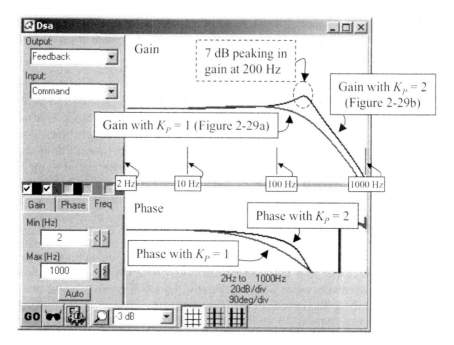

Figure 2-32. Output of DSA for $K_P = 1$ and $K_P = 2$.

of this discussion. However, it may be interesting to readers to notice that the two cases plotted in Figure 2-32 match the time-domain plots for Figures 2-29a and 2-29b, where the less stable Figure 2-29b corresponds to the plot in Figure 2-32 with peaking. Peaking and ringing are both indicators of inadequate margins of stability. Stability issues will be discussed in Chapter 3.

Like the Scope display, the DSA display is normally not visible when a model starts to run. The DSA display can be made visible by double-clicking inside the DSA block after the model has been compiled.

2.4.4.4 DSA Controls

The user can request a new Bode plot when the model is running by clicking on the *GO* button at bottom left of the DSA. This starts a new excitation period. For the experiments in this book, this process will continue for roughly 1 s of simulation time. After that, a new Bode plot will be displayed. Up to four plots can be saved. Right-click in the graph area of the DSA to bring up a pop-up menu and select *Save as* to save the most recent plot. Pressing the *GO* button a second time during the excitation period cancels the command for a new Bode plot.

To the right of the *GO* button, the sunglasses button hides the control panel. The gear button brings to view a dialog box for setting up the DSA excitation signal. The autofind button places a cursor according to the criteria in the adjoining combo box, which is set to 3 dB in Figure 2-32. The last three buttons control the number of cursors visible, allowing no cursors, one cursor, or two cursors.

2.4.4.5 The DSA Excitation Signal

The DSA works by generating a random command for a short period of time. The random signal is *rich*—it contains all the frequencies of the Bode plot. During the period of excitation, the DSA monitors all variables in the model. After the excitation, the DSA executes a fast Fourier transform (FFT) to convert the recorded data to a frequency-domain plot. When the random signal is applied to the model, the richness of the signal allows it to excite all frequencies at once. This is ideal for a modeling environment because it minimizes the time the DSA must excite the system. However, it also presents problems. First, the system must remain out of saturation—the power converter must not be driven beyond its maximum during the excitation. If a system is driven into saturation, the excitation amplitude can be reduced using the *Excitation Amplitude* node at the top left of the DSA (see Figure 2-31). However, if the amplitude is set too low, the signal-to-noise ratio of the system will be insufficient and the Bode plot will be distorted at high frequencies. Setting the amplitude of the excitation is sometimes a matter of experimentation. When doing so, always monitor the power converter output to ensure the system remains out of saturation for the entire excitation period. For all models in this book, the amplitude is set appropriately and users normally need not be concerned about this.

All commands except the DSA excitation must be shut off during the excitation period. The DSA will automatically disconnect the input node so that any signals connected to the input are disabled during DSA excitation; this is the case with the waveform generator in Figure 2-30. If there are waveform generators connected to other parts of the model, the *DSA Inactive* node at the lower right of Figure 2-31 can be wired to disable those generators. The *DSA Inactive* node is set to zero during the excitation period; when wired to a waveform generator *Enable* node, the desired behavior is realized.

2.4.5 Modeling Digital Control Systems

Experiment 2C, the final model of this chapter, will demonstrate how to model a simple digital control system in *Visual ModelQ*. This model, shown in Figure 2-33, is similar to Experiment 2B except that three blocks have been added. First, the PI controller, just below K_P and K_I, is now digital. The border area of this block is yellow in the *Visual ModelQ* environment and prints gray in the monochrome Figure 2-33.

Digital PI controllers sample the error at regular periods of time. The sample period for digital blocks is set via the controller node, the diamond at the bottom left of the PI block. The controller can be selected from multiple digital controllers, which

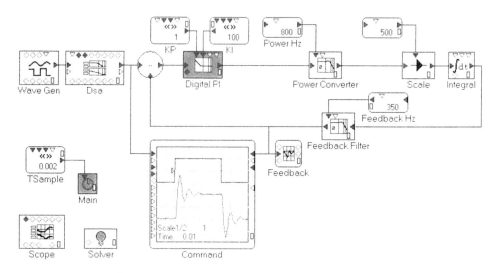

Figure 2-33. Experiment 2C: Experiment 2B with digital control.

can be running simultaneously in a *Visual ModelQ* model. Fortunately, most models are simple enough that one controller is sufficient. That controller is called *Main* in Experiment 2C and is near the center-left of Figure 2-33. The sole input node of the controller block is the sample time, which can be changed while the model is running. In Experiment 2C, that parameter is connected to a *Live Constant* named *TSample* to simplify changing the value.

Notice that the step response in Figure 2-33 overshoots and rings in Experiment 2C. All the parameters of Experiments 2B and 2C have identical defaults so one might have expected them to have a similar step response. Obviously, something is significantly different.

The difference between the two models is that Experiment 2C is the digital equivalent of Experiment 2B. The problem in Experiment 2C is that the sample time is too long for the dynamics of the system. As a result, the system is nearly unstable. Some experimentation can prove the point. Launch *Visual ModelQ* and load the file *Experiment_2C.mqd*. Click *Run*. Now, double-click on the *Live Constant* named *TSample*. Reduce the sample time by repeatedly clicking on the *Live Constant* "≪" button. When the sample time falls below about 0.0002 s, the response is equivalent to the analog performance. This is shown in Figure 2-34.

2.4.6 *Visual ModelQ* and This Book

This section has introduced several functions of *Visual ModelQ* used in this book. All key points of this book are demonstrated in *Visual ModelQ* models. Readers are encouraged to run these experiments and work the exercises at the end of each chapter.

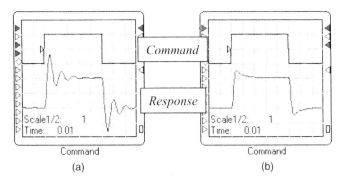

Figure 2-34. From Experiment 2C: Reducing sample time can stabilize a system. (a) TSample = 0.002 s; (b) TSample = 0.0002 s.

2.5 Exercises

1. Open *Experiment_2A.mqd* and click the *Run* button.
   A. Change the gain K_P from 1 to 2, and then raise it to 5. Describe what happens in the command response. What conclusion could you draw?
   B. Set $K_P = 2$ and change waveform to triangle. Are signs of marginal stability easier or harder to recognize? Repeat for sine wave and s-curve. What conclusion could you draw?
   C. Restore K_P to 1. Set K_I to 0. Describe what happens in the command response. Set K_I to a range of values from 10 to 1000. Describe what happens in the command response. What conclusion could you draw?
2. Open *Experiment_2B.mqd* and click the *Run* button.
   A. Run a Bode plot. Find the −3 dB frequency (the frequency where the gain falls to −3 dB) using the autofind combo box at the bottom of the DSA display window.
   B. Reduce control-loop gains. Set K_P to 0.5 and set K_I to 50. What is the gain at the frequency from 2A.
   C. Compare 2A and 2B. What conclusion could you draw?
3. Open *Experiment_2C.mqd* and click the *Run* button.
   A. Change *TSample* to several values spanning the range between 0.002 and 1×10^{-5} s. Over what range does the sample time significantly affect command response as viewed in the *Live Scope*?
   B. Does faster sampling make the system more stable or less stable?
   C. Set the sample time to 0.0001 s. Compare the step response of the digital system in Experiment 2C to the analog system in Experiment 2B. Repeat with $K_P = 2$. What conclusion could you draw?

Chapter 3

Review of the Frequency Domain

I n this chapter . . .

- Overview of the *s*-domain and the *z*-domain
- Detailed presentation of Mason's signal flow graphs
- Bode plots
- Measuring command response and stability
- The open-loop method
- A zone-based tuning procedure

This chapter will review the frequency domain, which is the basis for most analysis performed on control systems. The principles reviewed in this chapter are commonly taught in control-systems books, courses, and seminars so that many readers will find much of it familiar. In addition to the review, the final section provides a process for consistent tuning of controller gains; this process will be necessary to measure performance objectively, for example, when comparing traditional and observer-based systems. In addition, the same process will be applied to tuning observers in later chapters. For reference, most of this discussion is taken from [11, Chaps. 2–5].

3.1 Overview of the *s*-Domain

The Laplace transform underpins classic control theory [17, 37] and is defined in Equation 3.1 [7, p. 102] as

$$F(s) = \int_0^\infty f(t)e^{-st}dt, \qquad\qquad (3.1)$$

where $f(t)$ is a function of time, s is the Laplace operator, and $F(s)$ is the transformed function. The terms $F(s)$ and $f(t)$, commonly known as a *transform pair*, represent the same function in the two domains. For example, if $f(t)=\sin(\omega t)$, then $F(s)=\omega/(\omega^2+s^2)$. The Laplace transform moves functions between the time and the frequency domains. The most important benefit of the Laplace transform is that it provides s, the Laplace operator, and through that the frequency-domain transfer function.

3.1.1 Transfer Functions

Frequency-domain transfer functions describe the relationship between two signals as a function of s. For example, consider an integrator as a function of time. From Table 3-1, the integrator has an s-domain transfer function of $1/s$. So, it can be said for a system that produced an output, V_O, which was equal to the integral of the input, V_I, that:

$$\frac{V_O(s)}{V_I(s)} = \frac{1}{s}. \qquad\qquad (3.2)$$

The Laplace operator is a complex (as opposed to real or imaginary) variable. It is defined as

$$s \equiv \sigma + j\omega. \qquad\qquad (3.3)$$

The constant j is $\sqrt{-1}$. The ω term translates to a sinusoid in the time domain; σ translates to an exponential ($e^{\sigma t}$) term. The primary concern here is with steady-state sinusoidal signals, in which case $\sigma=0$. So in this book, σ will be ignored. To evaluate the DC response of a transfer function, set s to zero.

3.1.2 Linearity and the Frequency Domain

A frequency-domain transfer function is limited to describing elements that are linear and time invariant. These are severe restrictions and, in fact, virtually no real-world system fully meets them. The three criteria that follow define these attributes, the first two defining linearity and the third defining time invariance.

     1. *Homogeneity.* Assume that an input to a system $r(t)$ generates an output $c(t)$. For an element to be homogeneous, an input $k\times r(t)$ would have to generate an output $k\times c(t)$, for any value of k. An example of homogeneous behavior is an ideal resistor where $V=IR$. An example of nonhomogeneous behavior is saturation where twice as much input delivers less than twice as much output.

2. *Superposition.* Assume that an element, when subjected to the input $r_1(t)$ will generate the output $c_1(t)$. Further, assume that this same element, when subjected to the input $r_2(t)$ will generate the output $c_2(t)$. Superposition requires that if the element is subjected to the input, $r_1(t)+r_2(t)$, it will produce the output, $c_1(t)+c_2(t)$ [16, p. 93; 36].

3. *Time invariance.* Assume that an element has an input $r(t)$ that generates an output $c(t)$. Time invariance requires that $r(t-\tau)$ will generate $c(t-\tau)$ for all $\tau > 0$.

So the controls engineer faces a dilemma: transfer functions, the basis of classic control theory, require linear, time invariant (LTI) systems, but no real-world system is completely LTI. This is a complex problem that is dealt with in many ways. However, for most control systems, the solution is simple: design components close enough to being LTI that the non-LTI behavior can be ignored or avoided.

3.1.3 Examples of s-Domain Transfer Functions

Examples of transfer functions used in control laws are shown in Table 3-1. These functions can all be derived from Equation 3.1.

Integration and differentiation are the simplest operations. The s-domain operation of integration is $1/s$ and of differentiation is s. Filters are commonly used by control-systems designers such as when low-pass filters are added to reduce noise. Table 3-1 lists the s-domain representation for a few common examples. A compensator is a specialized filter: one that is designed to provide a specific gain and phase shift at one frequency. The effects on gain and phase either above or below that

TABLE 3-1 TRANSFER FUNCTIONS OF CONTROLLER ELEMENTS

Operation	Transfer function
Integration	$1/s$
Differentiation	s
Delay	e^{-sT}
Simple filters	
Single-pole low-pass filter	$K/(s+K)$
Double-pole low-pass filter	$\omega^2/(s^2+2\zeta\omega s+\omega^2)$
Notch filter	$(s^2+\omega^2)/(s^2+2\zeta\omega s+\omega^2)$
Bilinear-quadratic (bi-quad) filter	$(s^2+2\zeta\omega_N s+\omega_N^2)/(s^2+2\zeta\omega_D s+\omega_D^2)$
Compensators	
Lag	$K(\tau_Z s+1)/(\tau_P s+1),\ \tau_P>\tau_Z$
PI	$(K_I/s+1)K_P$
PID	$(K_I/s+1+K_D s)K_P$
Lead	$1+K_D s/(\tau_D s+1)\ or\ [(\tau_D+K_D)s+1]/(\tau_D s+1)$

frequency are secondary. Table 3-1 shows a lag compensator, a proportional-integral (PI) compensator, and a lead compensator.

Delays add time lag without changing amplitude. Since microprocessors have inherent delays for sampling, the delay function is especially important when analyzing digital controls. A delay of T seconds is defined in the time domain as

$$c(t) = r(t - T). \tag{3.4}$$

The corresponding function in the frequency domain is

$$T_{Delay}(s) = e^{-sT}. \tag{3.5}$$

3.1.4 Block Diagrams

Block diagrams are graphical representations developed to make control systems easier to understand. Blocks are marked to indicate transfer functions. In North America, transfer functions are usually indicated with their s-domain representation. The convention in Europe is to use schematic representation of a step response; Appendix C provides a listing of many North American and European block-diagram symbols.

Block diagrams can be simplified by combining blocks. Two blocks in parallel can be combined as their sum; two blocks in series can be represented as their product. When blocks are arranged to form a loop, they can be reduced using the $G/(1 + GH)$ rule. The forward path is $G(s)$ and the feedback path is $H(s)$. The transfer function of the loop is $G(s)/(1 + G(s)H(s))$ as shown in Figure 3-1.

The $G/(1 + GH)$ rule can be derived by observing in Figure 3-1a that the error signal ($E(s)$) is formed from the left as:

$$E(s) = R(s) - C(s) \times H(s).$$

$E(s)$ can also be formed from the right side (from $C(s)$ back through $G(s)$) as

$$E(s) = C(s)/G(s).$$

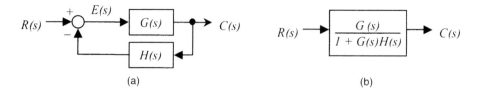

(a)   (b)

Figure 3-1. Simple feedback loop in equivalent forms.

So,

$$R(s) - C(s) \times H(s) = C(s)/G(s).$$

One or two steps of algebra produce:

$$\frac{C(s)}{R(s)} = \frac{G(s)}{1 + G(s)H(s)}.$$
(3.3b)

3.1.4.1 Mason's Signal Flow Graphs

An alternative to the $G/(1+GH)$ rule developed by Mason [10, p. 69; 7, p. 162; 28; 29; 31] provides graphical means for reducing block diagrams with multiple loops. The formal process begins by redrawing the block diagram as a signal flow graph.[1] The control system is redrawn as a collection of nodes and lines. Nodes define where three lines meet; lines represent the s-domain transfer function of blocks. Lines must be unidirectional; when drawn, they should have one and only one arrowhead. A typical block diagram is shown in Figure 3-2, and its corresponding signal flow graph is shown in Figure 3-3.

Step-by-step procedure. This section will present a step-by-step procedure to produce the transfer function from the signal flow graph based on Mason's signal flow graphs. The signal flow graph of Figure 3-3 will be used for an example. This graph has two independent inputs, $R(s)$ and $D(s)$. The example will find the transfer function from these two inputs to $D_O(s)$.

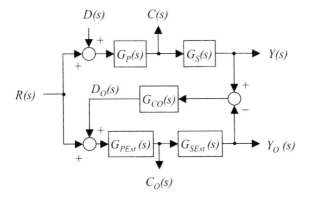

Figure 3-2. An example control-loop block diagram.

---

[1] For convenience, this step will be omitted in most cases and the block diagram will be used directly.

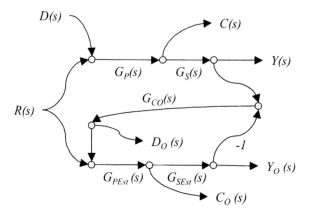

Figure 3-3. Signal flow graph for Figure 3-2.

Step 1: Find the loops. Locate and list all loop paths. For the example of Figure 3-3, there is one loop:

$$L_1 = -G_{PEst}(s) \times G_{SEst}(s) \times G_{CO}(s).$$

Step 2: Find the determinant of the control loop. Find the determinant, Δ, of the control loop, which is defined by the loops:

$\Delta = 1 - (\text{sum of all loops})$

$+ (\text{sum of products of all combinations of two nontouching loops})$

$- (\text{sum of products of all combinations of three nontouching loops})$

$+ \ldots.$

Two loops are said to be touching if they share at least one node. For this example there is only one loop:

$$\Delta = 1 + G_{PEst}(s) \times G_{SEst}(s) \times G_{CO}(s).$$

Step 3: Find all the forward paths. The forward paths are all the different paths that flow from the inputs to the output. For the example of Figure 3-3, there is one forward path from $D(s)$ to $D_O(s)$ and two from $R(s)$:

$$P_1 = D(s) \times G_P(s) \times G_s(s) \times G_{CO}(s)$$
$$P_2 = R(s) \times G_P(s) \times G_s(s) \times G_{CO}(s)$$
$$P_3 = R(s) \times G_{PEst}(s) \times G_{SEst}(s) \times -1 \times G_{CO}(s).$$

Step 4: Find the cofactors for each of the forward paths. The cofactor (Δ_K) for a particular path (P_K) is equal to the determinant (Δ) less loops that touch that path. For the example of Figure 3-3, all cofactors are 1 because every forward path includes $G_{CO}(s)$, which is in L_1, the only loop.

$$\Delta_1 = \Delta_2 = \Delta_3 = 1$$

Step 5: Build the transfer function. The transfer function is formed as the sum of all the paths multiplied by their cofactors, divided by the determinant:

$$T(s) = \frac{\sum_K (P_K \Delta_K)}{\Delta}. \tag{3.6}$$

For the example of Figure 3-3, the signal $D_O(s)$ is

$$D_O(s) = R(s)\frac{(G_P(s)G_S(s) - G_{PEst}(s)G_{SEst}(s))G_{CO}(s)}{1 + G_{PEst}(s)G_{SEst}(s)G_{CO}(s)} + D(s)\frac{G_P(s)G_S(s)G_{CO}(s)}{1 + G_{PEst}(s)G_{SEst}(s)G_{CO}(s)}.$$

Using a similar process, $C_O(s)$ can be formed as a function of $C(s)$ and $D(s)$:

$$C_O(s) = R(s)\frac{G_{PEst}(s)(1 + G_P(s)G_S(s)G_{CO}(s))}{1 + G_{PEst}(s)G_{SEst}(s)G_{CO}(s)} + D(s)\frac{G_{PEst}(s)G_P(s)G_S(s)G_{CO}(s)}{1 + G_{PEst}(s)G_{SEst}(s)G_{CO}(s)}.$$

As will be discussed in later chapters, a great deal of insight can be gained from transfer functions of this sort. Using Mason's signal flow graphs, transfer functions of relatively complex block diagrams can be written by inspection. Using the $G/(1+GH)$ rule to derive transfer functions from multiple-loop block diagrams will work but is more tedious.

It may be of interest that Figure 3-2 is an observer. The two blocks above represent the plant ($G_P(s)$) and the sensor ($G_S(s)$); those below are the approximations of those functions: the estimated plant, $G_{PEst}(s)$, and the estimated sensor, $G_{SEst}(s)$. The error between the actual and the estimated functions is fed through the observer compensator, $G_{CO}(s)$, which has high gains and will drive the output of the model (that is, the estimated plant and sensor) toward the output of the actual system. This form and its associated transfer functions will be the subject of the remaining chapters of this book.

3.1.5 Phase and Gain

The sine wave is unique among repeating waveforms; it is the only waveform that does not change shape when passing through LTI blocks. A sine-wave input generates a sine-wave output at the same frequency; the only differences possible between input and output are the gain and the phase. In other words, the response of an LTI system to any one frequency can be characterized completely, knowing only phase and gain.

Gain measures the difference in amplitude of input and output. Gain is often expressed in decibels or dB, which is defined as

$$\text{Gain} \equiv 20 * \text{Log}_{10}(\text{OUT}/\text{IN}), \tag{3.7}$$

where OUT and IN are the magnitudes of the output and input sine waves. This is shown in Figure 3-4. Phase describes the time shift between input and output. This lag can be expressed in degrees, where 360° is equivalent to one period of the input sine wave. Usually the output is considered as lagging (to the right of) the input. So, phase is defined as

$$\text{Phase} \equiv -360 \times F \times t_{DELTA}{}^{\circ}. \tag{3.8}$$

Frequently, the gain and phase of a transfer function are shown as a *phasor*, using the form *gain* dB∠–*Phase*°; for example, −3 dB∠−45°.

Phase and gain can be calculated from the transfer function by setting $s = j \times 2\pi f$, where F is the excitation frequency in Hertz. The response of a transfer function can be converted from a complex number to a phasor by using the magnitude of the number in decibels and taking the arc-tangent of the imaginary portion over the real, adding 180° if the real part is negative.

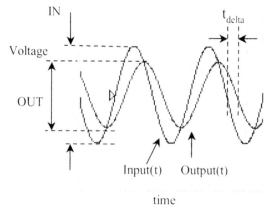

Gain = 20 Log$_{10}$(OUT/IN)   Phase = -360 × F × t_{delta}

Figure 3-4. Gain and phase.

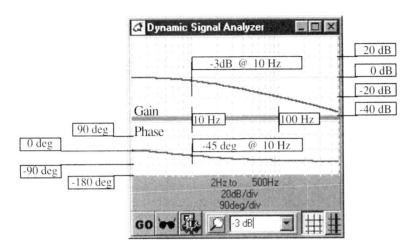

Figure 3-5. Typical Bode plot.

3.1.6 Bode Plots

Bode plots display phasors graphically; the gain in decibels and phase in degrees are plotted against the frequency in Hertz. The horizontal scale is logarithmic and the vertical scales are linear. Figure 3-5 shows a typical Bode plot. The frequency spans from 2 to 500 Hz (see legend just below the plot). The gain is shown in the top graph scaled at 20 dB per division with 0 dB at the solid center line. Phase is shown in the bottom graph scaled at 90° per division, again with 0° at the solid center line.

3.1.7 Measuring Performance

Objective performance measures provide a path for problem identification and correction. This section will discuss the two most common performance measures: command response and stability.

3.1.7.1 Experiment 3A: Measuring Performance

Experiment 3A (Figure 3-6) will be used to discuss performance issues. Experiment 3A is similar to Experiment 2B with two exceptions. First, the *Live Scope* has been enlarged slightly to show more detail. The two nodes at the bottom left of the block determine the size of the display. Second, a variable block *Error* has been added to give the DSA access to this signal; the DSA needs access to the loop error to support the open-loop method, which will be discussed later in this chapter.

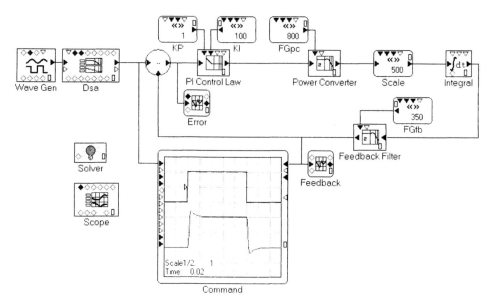

Figure 3-6. Experiment 3A, used to demonstrate performance measurement.

3.1.7.2 Command Response

Command response measures how quickly a system follows the command. In the time domain, the most common measure of command response is probably settling time to a step. True, few systems are subjected to step commands in normal operation. Still, the step response is useful because responsiveness can be measured on a scope more easily from a step than from most other waveforms.

Settling time is measured from the front edge of the step to the point where the feedback is within a certain value (typically between 1 and 10%) of the commanded value. If the response overshoots by more than the settling criterion, the system is considered settled only after the feedback has returned to within the settling criterion. For example, Figure 3-7a shows the step response of a control system to ±1 unit command. The excursion is 2 units so the system will be settled when the command is within 5% (0.1 units) of the final command or 1.0 ± 0.1 units. Since the feedback overshoots by more than 0.1 units, the system is settled well after the peak of the overshoot, when the feedback falls to 1.1 units. That time, shown in Figure 3-7a, is about 0.02 s.

Figure 3-7b shows a comparatively sluggish system; it requires 0.075 s to settle to 5%. To create a system with these characteristics, the control-law gains of Experiment 3A were reduced from their default values: K_P was reduced from 1.0 to 0.25 and K_I was reduced from 100 to 25. As expected, lower control-law gains produce less responsive performance.

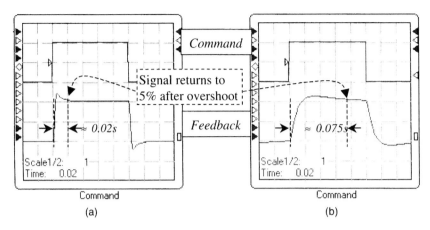

Figure 3-7. Step response of (a) responsive (K_P = 1, K_I = 100) and (b) sluggish systems. (K_P = 0.25, K_I = 25).

Response can also be measured in the frequency domain by inspecting the gain plot. Most control systems demonstrate good command response at low frequencies but become unresponsive at high frequencies. At low frequencies, the controller is fast enough to govern the system. As the frequency increases, the controller cannot keep up. Thinking of the transfer function, this means the gain at low frequencies will be approximately unity (1) but will be much less than unity at high frequencies. Consider a power supply that is advertised to produce sine-wave voltages up to 100 Hz. Such a power supply should be nearly perfect at low frequencies such as 1 Hz, begin to struggle as the frequency increases to its rated 100 Hz, and produce very low amplitude sinusoids at high frequencies, say, above 10 kHz. Translating this to a Bode plot, the gain would be about unity (0 dB) at low frequencies, start falling at mid-range frequencies, and continue falling to very low values at high frequencies. This is typical for control-system Bode plots.

Figure 3-8a shows the Bode plot for the system evaluated in Figure 3-7a. The frequency range spans from 2 to 1000 Hz. At low frequency, the gain is unity (0 dB). As the frequency increases, the gain rises a bit and then begins to fall. At the highest frequency shown, the gain has fallen more than two divisions, or −40 dB, which is equivalent to a gain of less than 1%. As the frequency increases, the gain will continue to fall. Closed-loop responsiveness is commonly measured in the frequency domain as the *bandwidth*, the frequency where the gain has fallen to −3 dB, or to a gain of about 70%. In Figure 3-8a, the bandwidth is about 180 Hz.

Figure 3-8b shows the Bode plot for the system of Figure 3-7b. The Bode plot for the sluggish system shows the bandwidth has fallen to 27.2 Hz. The bandwidth, like the settling time, shows the system in Figures 3-7a and 3-8a to be much more responsive.

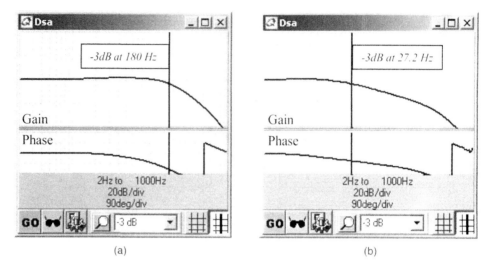

(a)   (b)

Figure 3-8. Frequency response (Bode plot) of (a) responsive and (b) sluggish systems.

3.1.7.3 Stability

Stability describes how predictably a system follows a command. In the time domain, stability is most commonly measured from the step response. The key characteristic is overshoot: the ratio of the peak of the response to the commanded change. The amount of overshoot that is acceptable in applications varies from 0 to perhaps 40%. Figure 3-9 shows the step response of two controllers. In Figure 3-9a, which is identical to the system from Figure 3-7a, the overshoot is a modest 20%.

A system with lower margins of stability is created in Figure 3-9b by changing the gains: K_P is reduced to 0.25 and K_I is increased to 250. In Figure 3-9b, the overshoot is more than 50%; worse, the overshoot is followed by ringing. Both systems are stable, but the margin of stability for the system on the right is too small for most applications.

Stability can also be measured from a system Bode plot; again the information is in the gain plot. As discussed above, at low frequencies, the gain will be at 0dB for most control systems and will fall off as the frequency increases. If the gain rises significantly before it starts falling, it indicates marginal stability. This phenomenon is called *peaking*. The amount of peaking is a measure of stability. For practical systems, allowable peaking ranges from 0 to perhaps 4dB. The two systems that were measured in the time domain in Figure 3-9 are measured again in Figure 3-10 using the frequency domain. Note Figure 3-10b with high peaking corresponds to the scope trace in Figure 3-9b with high overshoot.

The correlation between time and frequency domains can be seen by viewing the systems shown in Figures 3-7 through 3-10. Settling time correlates to bandwidth and overshoot correlates to peaking. For these simple systems, measures in either domain

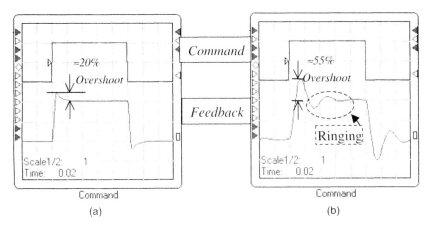

Figure 3-9. Step response of (a) stable ($K_P = 1$, $K_I = 100$) and (b) marginally stable ($K_P = 0.25$, $K_I = 250$) systems.

work well. The natural question is why both measures are needed. The answer is that in realistic systems the time domain is more difficult to interpret. Many phenomena in control systems occur in combinations of frequencies; for example, there may simultaneously be a mild resonance at 400 Hz and peaking at 60 Hz. Also, feedback resolution may limit the ability to interpret the response to small-amplitude steps. In real-world control systems, gleaning precise quantitative data from a step response is often impractical.

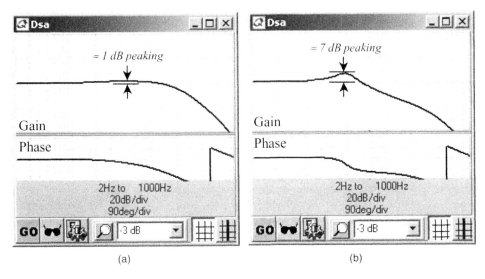

Figure 3-10. Frequency response (Bode plot) of (a) stable and (b) marginally stable systems.

Bode plots display effects at multiple frequencies with ease; correctly measured, Bode plots are less sensitive to resolution limitations. Thus, Bode plots can be relied upon to provide accurate measurements in real-world systems. However, time-domain plots are still often the preferred measurement because the equipment to make these measurements (chiefly, the oscilloscope) is readily available. Bode plots require a dynamic-signal analyzer (DSA), a device rarely found in laboratories.

3.2 Overview of the z-Domain

The z-domain is used to analyze digital controllers. The key feature of the z-domain is that it allows delays resulting from sampling to be accounted for easily. The s- and z-domains are so closely related that it may be a misnomer to refer to them as being in different domains. The basic principles — transfer functions, phase and gain, block diagrams, and Bode plots — are the same for both.

Digital controllers process data in regular intervals. The length of the intervals is referred to as T, the cycle or sample time. The z-domain is an extension of the s-domain [17]. It is based on the s-domain delay operation, which was shown in Equation 3.5 to be e^{-sT}. If $f(t)$ is delayed $N \times T$ seconds, then its Laplace transform is $e^{-sNT}F(s)$ or $(e^{-sT})^N F(s)$.

3.2.1 Definition of z

The term z is defined as e^{sT} [16, p. 127], which implies that $1/z$ is e^{-sT}, the delay operation. Developing the z-domain based on this simple equality may appear to require unwarranted effort. However, digital systems need to include the effects of delays so frequently that the effort is justified.

In the strictest sense, the s-domain is for continuous (not analog) systems and z is for sampled (not digital) systems. *Sampled* is synonymous with *digital* because digital systems normally are sampled; computers, the core of most digital controllers, cannot process data continuously. On the other hand, most analog systems are continuous. Recognizing this, *digital* in this book will imply *sampled* and *analog* will imply *continuous*.

3.2.2 z-Domain Transfer Functions

Transfer functions in z are similar to those in s in that both are usually ratios of polynomials. Several z-domain transfer functions are provided in Table 3-2. For example, consider a single-pole low-pass filter with a break frequency of 100 Hz (628 rad/s) and a sample time (T) of 0.001 s. From Table 3-2, the transfer function would be:

$$\frac{C(z)}{R(z)} = T(z) = \frac{z(1 - e^{-0.628})}{z - e^{-0.628}} = \frac{0.4663z}{z - 0.5337}. \tag{3.9}$$

TABLE 3-2 UNITY-DC-GAIN S-DOMAIN AND Z-DOMAIN FUNCTIONS

Operation	s-Domain transfer	z-Domain transfer function
Integration (accumulation)	$1/s$	$Tz/(z-1)$
Trapezoidal integration	$1/s$	$\dfrac{T}{2}\left(\dfrac{z+1}{z-1}\right)$
Differentiation (simple difference)	s	$(z-1)/Tz$
Inverse trapezoidal differentiation	s	$\dfrac{1+a}{T}\left(\dfrac{z-1}{z+a}\right),\ 0<a<1$
Delay T seconds	e^{-sT}	$1/z$
Simple filters ($\omega = 2\pi F$)		
Single-pole low-pass	$\omega/(s+\omega)$	$z(1-e^{-\omega T})/(z-e^{-\omega T})$
Two-pole low-pass[a]	$\omega_n^2/(s^2+2\zeta\omega_n s+\omega_n^2)$	$Az^2/(z^2+B_1 z+B_2)$ $B_1 = -2e^{-\zeta\omega_N T}\cos(\omega_N T\sqrt{1-\zeta^2})$ $B_2 = e^{-2\zeta\omega_N T}$ $A = 1+B_1+B_2$ $\zeta = Damping$
Two-pole notch[a]	$(s^2+\omega_n^2)/$ $(s^2+2\zeta\omega_n s+\omega_n^2)$	$K(z^2+A_1 z+A_2)/(z^2+B_1 z+B_2)$ $B_1 = -2e^{-\zeta\omega_N T}\cos(\omega_N T\sqrt{1-\zeta^2})$ $B_2 = e^{-2\zeta\omega_N T}$ $A_1 = -2\cos(\omega_N T)$ $A_2 = 1$ $K = (1+B_1+B_2)/(1+A_1+A_2)$ $\zeta = Damping$
Compensators		
PI	$(K_I/s+1)K_P$	$(K_I Tz/(z-1)+1)K_P$
Lead	$1+K_D s\cdot\omega/(s+\omega)$	$1+K_D(z-1)/Tz\cdot z(1-e^{-\omega T})/(z-e^{-\omega T})$

[a] If $\zeta > 1$, negate the term under the radical and substitute hyperbolic cosine for cosine.

The filters in Table 3-2 are developed by replacing polynomials of s-domain functions with the equivalent z-domain polynomials. For example, the term $(s-a)$ is replaced by $(z-e^{-aT})$; this process is repeated until all terms of s are replaced by z. Finally, a constant term is added so that the amplitude of the function at DC[2]$(z=1)$ is 1.

---

[2] The DC gain of a transfer function of z is evaluated by setting z to 1, which is equivalent to setting s to 0.

3.2.3 Bilinear Transform

An alternative to Table 3-2 for determining the z-domain equivalent of an s-domain transfer function is to approximate s as a function of z:

$$s \approx \frac{2}{T}\left(\frac{z-1}{z+1}\right). \tag{3.10}$$

This is called the *bilinear transformation* and is developed in Appendix D. This text will rely on Table 3-2 because it is usually less tedious than use of the bilinear transformation.

3.2.4 z Phasors

Phasors in z are similar to phasors in s. Again, the transfer function is evaluated with complex (versus real) math to determine the phase and gain at one frequency. The resulting complex number represents gain and phase as it did in the s-domain; the only difference is that z must be evaluated instead of s.

Evaluating z requires the following identity:

$$e^{jx} = \cos(x) + j \times \sin(x), \qquad \text{where } j = \sqrt{-1}.$$

Substituting z at steady state ($s = -j\omega$),

$$z = e^{+sT} = e^{+j\omega T} = \cos(\omega T) + j \times \sin(\omega T).$$

The magnitude of this equation is:

$$|z| = \sqrt{\cos^2(\omega T) + \sin^2(\omega T)} = 1.$$

And the angle is

$$\angle(z) = \tan^{-1}\left(\frac{\sin(\omega T)}{\cos(\omega T)}\right)$$
$$= \tan^{-1}(\tan(\omega T))$$
$$= \omega T.$$

This implies that the phasor representation for z is

$$z = e^{+sT}\big|_{s=j\omega} = 1\angle + \omega T. \tag{3.11}$$

The results of the equivalent filters in the s- and z-domains are similar but not identical. Digital functions are never identical to their analog counterparts, but they can be designed to be equivalent in a single facet of operation. A key advantage of the bilinear transformation is that it can be used to establish exact equivalence between the s-domain and the z-domain functions at one frequency; this requires the use of prewarping, a technique covered in Appendix D.

3.2.5 Bode Plots and Block Diagrams in z

Bode plots in the digital systems (that is, the z-domain) are the same as those in the s-domain. Block diagrams in the z- and s-domain are also the same. They can be combined with the $G/(1+GH)$ rule or Mason's signal flow graphs.

3.2.6 Sample-and-Hold

Digital controllers calculate the output once each cycle and hold it constant until the next cycle. This sample-and-hold (S/H) is present in virtually all digital systems. The effect of holding the output constant introduces phase lag because the output is aging from the time it is stored. At the start of the cycle the data is fresh, but by the end of the cycle, the output is a full cycle old. Since the stored data are, on average, one-half cycle old, the S/H acts approximately like a delay of a half cycle

$$T_{S/H}(z) \approx e^{-sT/2} = 1 \angle (-\omega T/2)\mathrm{rad} \qquad (3.12)$$

or, in degrees and Hz,

$$T_{S/H}(s) \approx 1 \angle (-180 \times F \times T)^\circ . \qquad (3.13)$$

At higher frequencies, the S/H also begins attenuating the input. The more exact transfer function for S/H is

$$T_{S/H}(z) = \frac{z-1}{Tz}\left(\frac{1}{s}\right), \qquad (3.14)$$

which is digital differentiation in series with analog integration. This form is shown as a zero-order hold in [7, p. 754], although the T is not included. Few textbooks include the T, although it is required to reflect the sample-and-hold's intrinsic unity DC gain. Recognizing that $z = e^{sT}$, some algebra can provide Equation 3.14 in a simpler form for sinusoidal excitation:

$$\begin{aligned}
T_{S/H}(z) &= \frac{z-1}{Tz}\left(\frac{1}{s}\right) \\
&= \frac{e^{sT}-1}{e^{sT}}\left(\frac{1}{Ts}\right) \\
&= \frac{e^{sT/2} - e^{-sT/2}}{e^{sT/2}}\left(\frac{1}{Ts}\right).
\end{aligned}$$

To apply steady-state sinusoids, set $s = j\omega$,

$$T_{S/H}(z) = \frac{e^{j\omega T/2} - e^{-j\omega T/2}}{e^{j\omega T/2}} \left(\frac{1}{Tj\omega} \right)$$

$$= e^{-j\omega T/2} \times \frac{e^{j\omega T/2} - e^{-j\omega T/2}}{2j} \times \frac{1}{\omega T/2}$$

$$= e^{-j\omega T/2} \times \frac{\sin(\omega T/2)}{\omega T/2}$$

$$= \left(\frac{\sin(\omega T/2)}{\omega T/2} \right) \angle (-\omega T/2) \text{rad.} \tag{3.15}$$

So, the precise sample-and-hold (Equation 3.15) and the approximation (Equation 3.12) have the identical phase lag, but different gains. The gain term, $\sin(\omega T/2)/(\omega T/2)$, also known as the *sync* function, is nearly unity for most frequencies of interest. For example, at one fourth the sample frequency ($\omega = 2\pi/4T$), the sync function evaluates to 0.9 dB, which is a value so close to 0 dB that it can usually be ignored. And recognizing that usually the system bandwidth will be at much lower frequencies, say one tenth the sample frequency, there is rarely much interest in the precise gain at so high a frequency. This is why the simpler Equation 3.12 is accurate enough to use in most control systems problems.

3.2.7 Quantization

Quantization is a nonlinear effect of digital control systems. It can come from resolution limitations of transducers or of internal calculations. One example is a 12-bit analog-to-digital converter (ADC) that converts a continuous range of 10 V to 4096 different values. Quantization also occurs in integer multiplications since the resultant of a multiplication is usually rounded.

If the resolution of sensors and the control-system mathematics is fine enough, quantization can be ignored. Otherwise it must be taken into account either in modeling or by use of statistical methods. Quantization is nonlinear and cannot be represented in the z-domain. One effect of quantization is called limit cycles [2, p. 367; 16]. Limit cycles are low-level oscillations that occur because of quantization error in digital mathematics. Limit cycles can produce sustained oscillations that are low in frequency and many times larger than the quantization level.

An ADC converts a voltage to an integer, where the value of the integer is proportional to the amount of voltage. Similarly, digital-to-analog converters (DACs) convert integers to voltages. A sample-and-hold is an implicit part of the output DAC. That is, a DAC can be modeled with two sections: a constant in volts per bit and a S/H. In general, the model for DACs and ADCs is the ratio of the integer range and the voltage range. When studying the effects of phase lag on stability, the sample-and-hold usually can be placed anywhere in the loop.

3.3 The Open-Loop Method

The *open-loop method* is a technique that analyzes margins of stability. It simplifies the complex task of evaluating the stability of a control system to calculations based on the phase and gain at two key frequencies. Many competing methods of stability analysis, such as the popular *root locus method*, rely on subjective measures of stability margins such as evaluating graphical patterns within complex plots. Such methods are often impractical in realistic systems because the graphs grow increasingly difficult to interpret as system models are augmented to take into account more detailed behavior of the system. The open-loop method allows any known LTI behavior to be accounted for, and it does so without significantly increasing the difficulty of executing the method.

The open-loop method simplifies any control loop to the diagram shown in Figure 3-11. All components in the forward path, including the control law, the power converter, and the plant, are combined to form $G(s)$. $H(s)$ includes all components in the feedback path.

The closed-loop transfer function is the relationship between the command, $R(s)$, and plant output, $C(s)$, which is equivalent to $G/(1+GH)$. (For convenience, the closed-loop function is often taken substituting $F(s)$ for $C(s)$ because $C(s)$ cannot be measured directly in physical systems.) The open-loop transfer function is defined as the loop gain, which is the path from $E(s)$ to $F(s)$. This is equivalent to GH. Figure 3-12 shows a closed-loop ($G/(1+GH)$) and Figure 3-13 shows the open-loop (GH) transfer function for the model of Experiment 3A with default parameters.

The open-loop method provides two margins of stability, *gain margin* and *phase margin*. Both are based on the understanding that instability occurs when the open-loop gain is unity (0 dB) and the phase is −180°. This can be seen with the $G/(1+GH)$ rule which reduces to $G/0$ when $GH=0$ dB $\angle-180°$ since 0 dB $\angle-180°$ is equivalent to −1. The open-loop method measures by how much of a margin the control system avoids instability.

The *Visual ModelQ* DSA in Experiment 3A is configured to make convenient the display of both the open-loop and the closed-loop plots. The closed loop is the relationship between the command and feedback signals. Accordingly, the input and

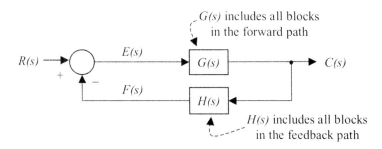

Figure 3-11. The open-loop method simplifies the control loop to two transfer functions.

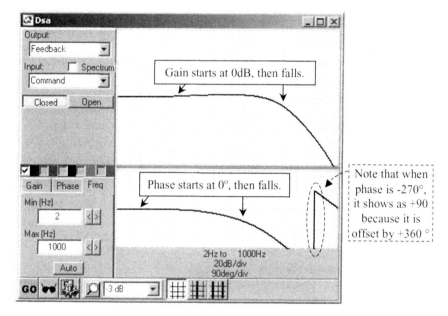

Figure 3-12. Characteristics of closed-loop Bode plot on a stable system.

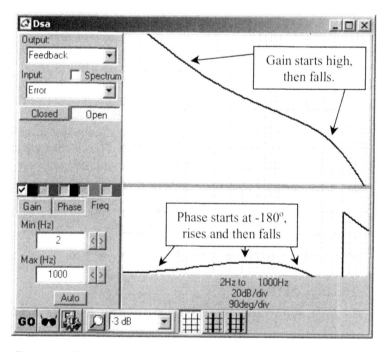

Figure 3-13. Characteristics of an open-loop Bode plot corresponding to Figure 3-12.

output channels of the DSA are selected as *Command* and *Feedback* in the upper left of Figure 3-12. The open loop describes the path around the loop. Referring to Figure 3-6, this path starts at the variable block *Error*, which is just right of the summing junction, and follows the loop back around to the *Feedback* variable block. Thus, the DSA in Figure 3-13 shows the input channel for the open-loop plot as *Error*. The DSA in Experiment 3A is configured to have two buttons, *Closed* and *Open*, which select the appropriate input and output channels for the two plots. The DSA buttons are configured with the button setup node at the top right of the DSA block.

Phase margin or PM [4, 36] is the margin of stability measured at the frequency where the gain around the loop falls to 0 dB. As shown in Figure 3-13, control systems usually have a large open-loop gain (≫0 dB) at low frequencies and a small loop gain (≪0 dB) at high frequencies. For at least one frequency, the gain will pass through 0 dB; that frequency is called the *gain-crossover frequency*. Were the phase equal to −180° at that frequency, the system would be unstable. The further the phase is from −180°, the greater the margin of stability. The PM is defined as the difference of the actual phase and −180°. Figure 3-14 shows the open-loop plot of Figure 3-13 with the gain-crossover frequency and PM identified as 81 Hz and 52°, respectively.

Gain margin or GM is the margin of stability measured at the frequency where the phase around the loop falls to −180°. As shown in Figure 3-14, the loop phase will

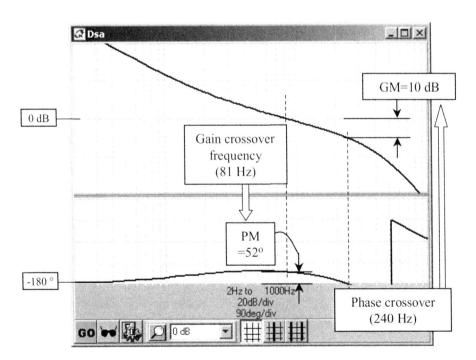

Figure 3-14. Open-loop Bode plot showing how to calculate phase and gain margins.

usually be greater than −180° at the gain-crossover frequency and start falling at higher frequencies. At one frequency, the gain will pass through −180°; that frequency is called the *phase-crossover frequency*. Were the gain equal to 0 dB at that frequency, the system would be unstable. The further the gain is from 0 dB, the greater the margin of stability. The GM is defined as the negative of the actual gain at the phase crossover. The system in Figure 3-14 identifies the GM as 10 dB and the phase-crossover frequency as 240 Hz.

3.3.1 Specifying the Target GM and PM

Although the measurement of PM and GM is objective, determining the desired values for these measures requires judgment. One reason is that applications vary in the amount of margin they require. For example, some applications must follow commands like the step that will generate overshoot in all but the most stable systems. These applications often require higher margins of stability than those systems that respond only to gentler commands. Also, some applications can tolerate more overshoot than others. Finally, some control laws require more PM or GM than others for equivalent response. For example, a PI controller requires typically about 55° of PM to achieve 20% overshoot to a step, while a PID (proportional-integral-differential) controller might be able to eliminate all overshoot with just 40° PM.

In practice, the GM of a well-tuned system will fall between 6 and 20 dB, depending on the application and the controller type; PM will fall between 35° and 80°. All things being equal, more PM is better. This teaches one of the most basic rules in controls: *Eliminate unnecessary phase lag!* Every noise filter, feedback device, and power converter contributes to phase lag around the loop and each erodes the PM. Unnecessary lags limit the ultimate performance of the control system.

3.4 **A Zone-Based Tuning Procedure**

One challenge of tuning is that multiple gains must be varied and each affects many of the performance measures. The goal of the *zone-based* tuning process is to tune the multiple tuning gains one at a time. This can be done by considering the effects of each gain term as being dominant over a certain zone of frequency [25, 26]. A PI controller has two zones, one for the proportional term and the other for the integral term. Consider the PI control loop in Figure 3-15.

The frequency zones are easiest to see when the effects of the power converter and feedback filter are ignored. So, setting those two blocks to 1 and using the $G/(1+GH)$ rule with some algebra, the simplified closed-loop transfer function is

$$\frac{C(s)}{R(s)} = \frac{sK_P + K_I K_P}{s^2/G + sK_P + K_I K_P}. \tag{3.16}$$

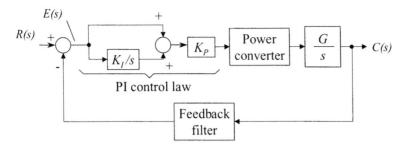

Figure 3-15. A simplified PI controller.

To consider (3.16) in terms of frequency zones, focus on the denominator. At high frequencies s is large and s^2/G dominates the denominator. This is intuitive; at very high frequencies, almost all control systems respond according to the plant gain (inductance, capacitance, inertia, etc.); at those frequencies, the response is beyond the reach of the controller.

As the frequency declines, the $s \times K_P$ term will become dominant. This is the *middle-frequency* range. In the control law, the proportional gain dominates; the frequency is still too high for the integral gain to have much effect. As the frequency declines further, the integral term will dominate because a low value of the s terms will leave only the constant $K_I K_P$ term in the numerator and denominator of Equation 3.16. This is the *low-frequency* zone.

Understanding the concept of frequency zones, a tuning procedure can be developed where one gain is tuned at a time. This eliminates the problems associated with tuning multiple gains simultaneously. The method that follows assumes the user has only time-domain measurements, as this is common in industry. The frequency-domain plots are shown for reference.

3.4.1 Zone One: Proportional

To apply the zone-based approach to tuning, tune the highest frequency term first. Assuming the plant gain, G, cannot be changed, start by tuning K_P. If the application can tolerate it, set K_I to zero and set K_P very low. Note that some applications cannot tolerate the lack of an integral such as in motion control, when a velocity loop is applied to a vertical load (the load can fall). Apply a square wave and raise K_P until significant overshoot is generated. How much overshoot is significant depends on the application. Many applications can tolerate 10 or 15%, while others can tolerate none at all. For applications that demand high responsiveness, it is important to raise K_P as high as possible. In Figure 3-16, a very small amount of overshoot is assumed to be tolerable and the result is $K_P = 1$.

By looking at the open-loop gain, the advantage of setting K_I to 0 when tuning K_P is apparent. Notice that at low frequencies the open-loop phase is $-90°$ compared to

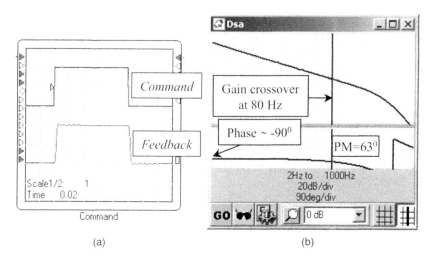

(a)   (b)

Figure 3-16. From Experiment 3A: Tuning Zone 1 using (a) step response and (b) open-loop plot of P-controller for $K_P = 1$, $K_I = 0$.

the $-180°$ seen in the PI controller of Figure 3-14. So, when $K_I = 0$, no matter how low K_P is set, the system will not become unstable. In the PI controller of Figure 3-14, when the gain crossover frequency is well above or well below the phase "hump" (about 50 Hz), the PM falls so that the system is not sufficiently stable. One advantage of starting with a P controller is that the designer can start with K_P very low and have little risk of instability.

A natural question is what causes instability seen with high K_P? As always, instability is caused by phase lag accumulating to $-180°$ with 0 dB gain. Phase from the plant here is fixed at $-90°$, but phase lag from the power converter and feedback filter increases as frequency rises; this accounts for decline in open-loop phase after the gain-crossover frequency shown in Figure 3-16b. K_P is raised to provide the maximum responsiveness possible within a minimum margin of stability. If the maximized K_P is insufficient, the designer must look toward reducing phase lag in the components of the loop, for example, by increasing the responsiveness of the power converter or the feedback device.

3.4.2 Zone Two: Integral

After K_P is maximized, attention should be turned to the next zone, which is served by the integral term. Integral gain is important because it removes the long-term error. When tuning, the value of integral gain will have little bearing on the stability of the next higher frequency (K_P) zone; the implication is that changing K_I will not require returning to change K_P. Continuing from above, raise K_I from zero

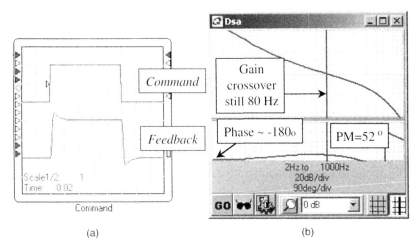

Figure 3-17. (a) Step response and (b) open-loop plot of PI controller with $K_I = 100$, $K_P = 1$.

until overshoot is excessive — usually between 10 and 30%. For example, the overshoot is about 10% when $K_I = 100$ in Figure 3-17a; the open-loop Bode plot is shown in Figure 3-17b.

Notice in Figure 3-17b that the gain crossover remained at 80 Hz, but the PM dropped 10°. Both are expected; the integral did not change the crossover frequency because its frequency zone is well below 80 Hz. So, K_I has little effect on gain at the gain-crossover frequency, but it still contributes 10° of phase lag at that frequency, reducing the PM by that same amount. Notice that the PI controller has overshoot. All PI controllers generate some overshoot in response to a square wave; it is one of the weaknesses of the method.

The zone-based procedure shown here can be extended to more complicated controllers. For example, if a double integral is added to the control law, it becomes a third zone, which becomes the lowest frequency zone and is tuned after the first two zones are tuned. The key is to divide the control laws into multiple zones and then tune the terms, starting with the highest zone and moving to the lowest.

One exception to this rule is when a PID controller is used to regulate a single-integrating plant. In this case, the proportional and derivative terms form a single zone. When tuning such a controller, zero the integral and derivative gains and tune the proportional term for lower-than-normal margins of stability, allowing perhaps 10 or 20% more overshoot than will be acceptable in the final tuning. Then add a small amount of derivative gain to increase the stability margins. In this way, the derivative and proportional gains combine to form the highest frequency zone. The integral zone, which is the next lower zone, is tuned the same way as it would be in the PI controller.

3.5 Exercises

1. Confirm that the zone-based method provides consistent margins of stability. Open *Experiment_3A.mqd* and click the *Run* button.
   A. Set the feedback filter frequency (*FGfb*) to 200 and retune the system using the zone-based procedure of this chapter; do not allow any overshoot with K_P and allow only 10% overshoot with K_I. Measure PM and GM.
   B. Repeat for *FGfb* = 350.
   C. Repeat for *FGfb* = 500.
   D. Repeat for *FGfb* = 1000.
   E. Relate the time-domain data of parts A–D to GM and PM for the system of *Experiment_3A.mqd*.
   F. How could you modify the procedure to generate somewhat lower margins of stability?

2. Closed-loop bandwidth.
   A. Measure the closed-loop bandwidth (−3 dB frequency) for each of the cases in 1A–1D.
   B. What is the relationship between low-pass filters in the loop and maximum command response?

3. Create a zone-based tuning procedure that produces tuning gains in Experiment 3A so margins of stability are GM = 10 dB and PM = 50°.
   A. What tuning values (K_P, K_I) are produced for *FGfb* = 200?
   B. Repeat for *FGfb* = 350.
   C. Repeat for *FGfb* = 500.
   D. Repeat for *FGfb* = 1000.

4. Closed-loop bandwidth.
   A. Measure the closed-loop bandwidth (−3 dB frequency) for each of the cases in 3A–3D.
   B. Compare bandwidths in 4A to those in 2A. What is the relationship between more aggressive tuning (lower PM and GM) and maximum command response?

Chapter 4

The Luenberger Observer: Correcting Sensor Problems

I n this chapter . . .

- Development of the Luenberger observer
- Experiments demonstrating how observers enhance stability
- Practical aspects of designing and tuning observers

This chapter will introduce the Luenberger observer. It will focus on the *predictor–corrector* structure where the observer uses a model to predict system operation and then uses the feedback signal to correct deviations between the model and the actual system. Also, a filter form of the observer will be discussed; this representation provides additional insight into the operation of observers. This chapter will conclude with a step-by-step procedure for designing a Luenberger observer.

4.1 What Is a Luenberger Observer?

An observer is a mathematical structure that combines sensor output and plant excitation signals with models of the plant and sensor. An observer provides feedback signals that are superior to the sensor output alone. The topic of this book is the Luenberger observer, which combines five elements:

- a sensor output, $Y(s)$,
- a power converter output (plant excitation), $P_C(s)$,
- a model (estimation) of the plant, $G_{PEst}(s)$,

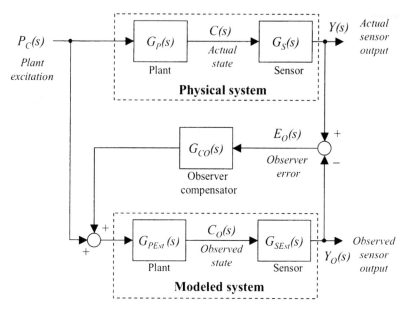

Figure 4-1. General form of the Luenberger observer.

- a model of the sensor, $G_{SEst}(s)$, and
- a PI or PID observer compensator, $G_{CO}(s)$.

The general form of the Luenberger observer is shown in Figure 4-1.

4.1.1 Observer Terminology

The following naming conventions will be used. *Estimated* will describe components of the system model. For example, the estimated plant is a model of the plant that is run by the observer. *Observed* will apply to signals derived from an observer; thus, the state (C_O) and the sensor (Y_O) signals are *observed* in Figure 4-1. Observer models and their parameters will be referred to as *estimated*. Transfer functions will normally be named $G(s)$ with identifying subscripts: $G_P(s)$ is the plant transfer function and $G_{PEst}(s)$ is the estimated or modeled plant.

4.1.2 Building the Luenberger Observer

This section describes the construction of a Luenberger observer from a traditional control system, adding components step by step. Start with the traditional control system shown in Figure 4-2. Ideally, the control loop would use the actual state, $C(s)$, as feedback. However, access to the state comes through the sensor, which produces

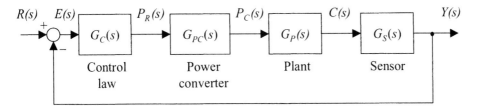

Figure 4-2. Traditional control system.

$Y(s)$, the feedback variable. The sensor transfer function, $G_S(s)$, often ignored in the presentation of control systems, is the focus here. Typical problems caused by sensors are phase lag, attenuation, and noise.

Phase lag and attenuation can be caused by the physical construction of the sensor or by sensor filters, which are often introduced to attenuate noise. The key detriment of phase lag is the reduction of loop stability. Noise can be generated by several forms of electromagnetic interference (EMI). Noise causes random behavior in the control system, corrupting the output and wasting power. All of these undesirable characteristics are represented by the term $G_S(s)$ in Figure 4-2. The ideal sensor can be defined as $G_{S\text{-}IDEAL}(s)=1$.

The first step in dealing with sensor problems is to select the best sensor for the application. Compared to using an observer, selecting a faster or more accurate sensor will provide benefits that are more predictable and more easily realized. However, limitations such as cost, size, and reliability will usually force the designer to accept sensors with undesirable characteristics, no matter how careful the selection process. The assumption from here forward will be that the sensor in use is appropriate for a given machine or process; the goal of the observer is to make the best use of that sensor. In other words, the first goal of the Luenberger observer will be to minimize the effects of $G_S(s)\neq1$.

For the purposes of this development, only the plant and sensor, as shown in Figure 4-3, need to be considered. Note that the traditional control system ignores the effect of $G_S(s)\neq1$; $Y(s)$, the sensor output, is used in place of the actual state under control, $C(s)$. But $Y(s)$ is not $C(s)$; the temperature of a component is not the temperature indicated by the sensor. Phase lag from sensors often is a primary contributor to loop instability; noise from sensors often demands correction by the addition of filters in the control loop, again contributing phase lag and ultimately reducing margins of stability.

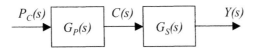

Figure 4-3. Plant and sensor.

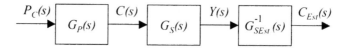

Figure 4-4. An impractical way to estimate $C(s)$: Adding the inverse sensor transfer function.

4.1.2.1 Two Ways to Avoid $G_S(s) \neq 1$

So, how can the effects of $G_S(s) \neq 1$ be removed? One alternative is to follow the sensed signal with the inverse of the sensor transfer function: $G_{SEst}^{-1}(s)$. This is shown in Figure 4-4. On paper, such a solution appears workable. Unfortunately, the nature of $G_S(s)$ makes taking its inverse impractical. For example, if $G_S(s)$ were a low-pass filter, as is common, its inverse would require a derivative as shown in Equation 4.1. Derivatives are well known for being too noisy to be practical in most cases; high-frequency noise, such as that from quantization and EMI, processed by a derivative generates excessive high-frequency output noise.

$$\text{If } G_{SEst}(s) = \frac{K_{EST}}{s + K_{EST}}, \quad \text{then } G_{SEst}^{-1}(s) = 1 + \frac{s}{K_{EST}}. \qquad (4.1)$$

Another alternative to avoid the effects of $G_S(s) \neq 1$ is to simulate a model of the plant in software as the control loop is being executed. The signal from the power converter output is applied to a plant model, $G_{PEst}(s)$, in parallel with the actual plant. This is shown in Figure 4-5. Such a solution is subject to drift because most control-system plants contain at least one integrator; even small differences between the physical plant and the model plant will cause the estimated state, $C_{Est}(s)$, to drift. As a result, this solution is also impractical.

The solution of Figure 4-4, which depends wholly on the sensor, works well at low frequency but produces excessive noise at high frequency. The solution of Figure 4-5, which depends wholly on the model and the power converter output signal, works well at high frequency but drifts in the lower frequencies. The Luenberger observer, as will be shown in the next section, can be viewed as combining the best parts of these two solutions.

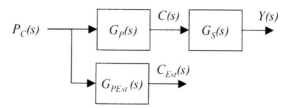

Figure 4-5. Another impractical solution: Deriving the controlled state from a model of the plant.

4.1.2.2 Simulating the Plant and Sensor in Real Time

Continuing the construction of the Luenberger observer, augment the structure of Figure 4-5 to run a model of the plant and sensor in parallel with the physical plant and sensor. This configuration, shown in Figure 4-6, drives a signal representing the power conversion output through the plant model and through the sensor model to generate the observed sensor output, $Y_O(s)$. Assume for the moment that the models are exact replicas of the physical components. In this case, $Y_O(s) = Y(s)$, or, equivalently, $E_O(s) = 0$. In such a case, the observed state, $C_O(s)$, is an accurate representation of the actual state. So $C_O(s)$ could be used to close the control loop; the phase lag of $G_S(s)$ would have no effect on the system. This achieves the first goal of observers, the elimination of the effects of $G_S(s) \neq 1$, but only for the unrealistic case where the model is a perfect representation of the actual plant.

4.1.2.3 Adding the Observer Compensator

In any realistic system $E_O(s)$ will not be zero because the models will not be perfect representations of their physical counterparts and because of disturbances. The final step in building the Luenberger observer is to route the error signal back to the model to drive the error toward zero. This is shown in Figure 4-7. The observer compensator, $G_{CO}(s)$, is usually a high-gain PI or PID control law.

The gains of $G_{CO}(s)$ are often set as high as possible so that even small errors drive through the observer compensator to minimize the difference between $Y(s)$ and $Y_O(s)$. If this error is small, the observed state, $C_O(s)$, becomes a reasonable representation of the actual state, $C(s)$; certainly, it can be much more accurate than the sensor output, $Y(s)$.

One application of the Luenberger observer is to use the observed state to close the control loop; this is shown in Figure 4-8, which compares to the traditional control system of Figure 4-2. The sensor output is no longer used to close the loop; its sole function is to drive the observer to form an observed state. Typically, most of the

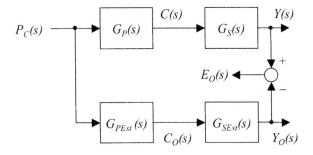

Figure 4-6. Running models in parallel with the actual components.

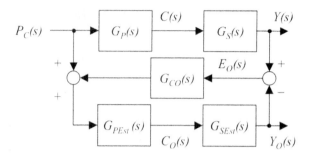

Figure 4-7. The Luenberger observer.

phase lag and attenuation of the sensor can be removed, at least in the frequency range of interest for the control loop.

4.2 Experiments 4A–4C: Enhancing Stability with an Observer

Experiments 4A–4C are *Visual ModelQ* models that demonstrate one of the primary benefits of Luenberger observers: the elimination of phase lag from the control loop and the resulting increase in margins of stability. Experiment 4A represents the traditional system. Experiment 4B restructures the loop so the actual state is used as the feedback variable. Of course, this is not practical in a working system (the actual state is not accessible) and is only used here to demonstrate the negative effects of the sensor. Experiment 4C adds a Luenberger observer to the control system. The result will be that the system performance will be equal to that of Experiment 4B where the

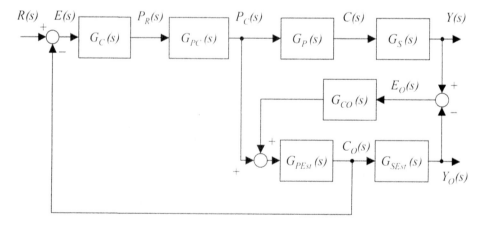

Figure 4-8. A Luenberger observer-based control loop.

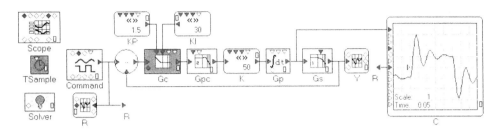

Figure 4-9. Experiment 4A: A traditional control system.

actual state was used. In other words, the effects of the imperfect sensor will be eliminated by the observer.

The model of Experiment 4A is shown in Figure 4-9. The loop includes a PI compensator, a 50-Hz bandwidth power converter, and a 20-Hz bandwidth sensor. The PI compensator is tuned aggressively so that the margins of stability are too low. The sensor is the primary contributor to phase lag in the control loop and thus the primary cause of low margins of stability. Starting from the left, there are ten components in this model:

Command	A waveform generator used to excite the system with a 2-Hz square wave at an amplitude of ±1.
R	A variable block for R, the command. R is not used in this discussion; it and several other variables are provided for readers who wish to explore the model independently. To view these variables, run the model and double-click on the main scope block to see the main scope display, which will display all variables.
Subtraction	A subtraction block to form the error, $R - Y$.
G_C	A digital PI control law operating on the error signal. The PI gains are set high enough to cause overshoot and ringing: $K_I = 30.0$ and $K_P = 1.5$.
G_{PC}	The power converter modeled by a two-pole low-pass filter with a bandwidth of 50 Hz and a ζ of 0.707.
K	A scaling gain representing part of the plant transfer function. The gain here is 50.
G_P	An integrator that, together with K, represents the plant.
G_S	A single-pole low-pass filter representing the sensor. The bandwidth of the filter is 20 Hz. This bandwidth is so low that the sensor is the predominant source of phase lag in the control loop.
Y	A variable block for Y, the system output.
C	A *Live Scope* for C, the actual state.

The model includes a *Live Scope* display of C, the actual state. This signal is the system response to R, the square-wave command. The signal C shows considerable

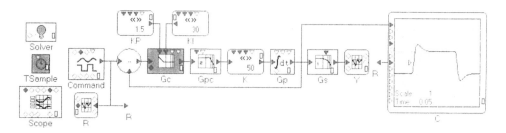

Figure 4-10. From Experiment 4B: An idealized system, which uses the actual state for feedback, has conservative margins of stability.

overshoot and ringing, which indicates marginal stability. The gains of G_C, the PI controller ($K_P = 1.5$ and $K_I = 30$), are set too aggressively for the phase lag of this loop.

The marginal stability of Figure 4-9 is caused in large part by the phase lag created by G_S. Were G_S an ideal sensor ($G_S = 1.0$), the system would behave well with the gains used in that model. This is demonstrated in Experiment 4B (see Figure 4-10), which is identical in all respects to Figure 4-9 except that the loop feedback is the actual state (C), not the measured output (Y). Of course, this structure is impractical: loops must be closed on measured signals. In traditional control systems, the use of sensors with significant phase lag[1] usually requires reducing control-law gains, which implies a reduction in system performance. In other words, the most common solution for the problems of Figure 4-9 is to reduce the PI gains and accept the lower level of performance. An observer can offer a better alternative.

Experiment 4C (see Figure 4-11) is Experiment 4A with an observer added. Several new blocks are used to construct the observer of Figure 4-7:

Subtraction A subtraction block is added on the right to take the difference of Y, the actual feedback signal, and Y_o, the observed feedback signal. This forms the observer error signal.

G_{CO} A PID control law is used as the observer compensator. G_{CO} was tuned experimentally using a process that will be described later in this chapter. G_{CO} samples at 1 kHz, as do the other digital blocks (Delay, G_{PEst}, and G_{SEst}).

Addition An addition block is added to combine the output of the observer compensator and the power converter output.

K_{Est} An estimate of the gain portion of the plant. Here, the gain is set to 50, the gain of the actual plant ($K_{Est} = K$).

---

[1] Significant in comparison to the other sources of phase lag in the loop such as internal filters and the power converter.

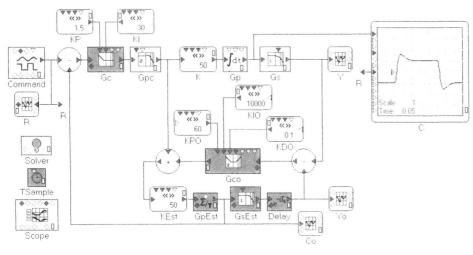

Figure 4-11. Model of observer-based control system, from Experiment 4C.

G_{PEst}	A digital integrator, which is a summation scaled by $1/T$. It is used to estimate the second part of the plant, G_P.
G_{SEst}	A digital filter used to model the sensor transfer function. This is a single-pole, low-pass filter with a bandwidth of 20 Hz. This is an accurate representation of the sensor.
C_O	Variable block C_O, the observed state.
Delay	A delay of one sample time. This delay recognizes that there must be a sample–hold delay at some position in the digital-observer loop. In this case, it is assumed that during each cycle of the observer, Y_O is calculated to be used in the succeeding cycle.

The other significant change in Experiment 4C is that the feedback for the control loop is taken from the observed state, not the sensor output, as was the case in Experiment 4A. The results of these changes are dramatic. The actual state shown in Figure 4-11 shows none of the signs of marginal stability that were evident in Figure 4-9. This improvement is wholly due to reduction of phase lag from the sensor. In fact, the response is so good, it is indistinguishable from the response of the system closed using the actual state as shown in Figure 4-10 (the desirable, albeit impractical, alternative).

Here, the Luenberger observer provided a practical way to eliminate the effects of sensor phase lag in the loop. Further, these benefits could be realized with a modest amount of computational resources: a handful of simple functions running at 1 kHz. Altogether, these calculations represent a few microseconds of computation on a modern DSP and not much more on a traditional microprocessor. In many cases, the calculations are fast enough that they can be run on existing hardware using uncommitted computational resources.

4.2.1 Experiment 4D: Elimination of Phase Lag

A brief investigation can verify the elimination of phase lag demonstrated with Experiments 4A–4C. Experiment 4D (see Figure 4-12) displays three signals on *Live Scopes*: *C*, the actual state, *Y*, the measured state, and C_O, the observed state. At first glance, all three signals appear similar. However, upon closer inspection, notice that *Y* (rightmost) lags *C* (top) by about a division. For example, *C* crosses two vertical divisions at two time divisions after the trigger time, which is indicated by a small triangle; *Y* requires three time divisions to cross the same level, about 10 ms longer. Note that the time scale has been reduced to 0.01 s in Experiment 4D to show this detail. Note also that all three *Live Scopes* are triggered from *R*, the command, and so are synchronized.

Now compare the observed state, C_O, to the actual state. These signals are virtually identical. For example, C_O crosses through two vertical divisions at the second time division, just like the actual state. The phase lag from the sensor has been eliminated by the observer. This explains why the margins of stability were the same whether the system used the actual state (Experiment 4B) or the observed state (Experiment 4C) for feedback and why the margins were lower in the system using the measured state (Experiment 4A).

It should be pointed out that this observer has several characteristics that will not be wholly realized in a practical system: the plant and sensor models are near-perfect

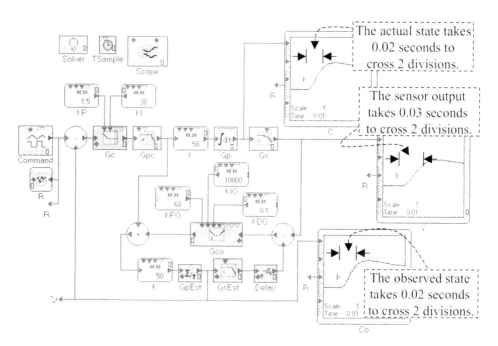

Figure 4-12. Experiment 4D: Three signals in an observer-based system.

representations of the actual plant and system, there are no disturbances, and the sensor signal is noiseless. Certainly, these imperfections will limit the benefits of observers in real-world control systems. Even so, the principle shown here is reliable in working machines and processes: the Luenberger observer can produce substantial benefits in the presence of sensor phase lag, and it can do so with modest computational resources.

4.3 Predictor–Corrector Form of the Luenberger Observer

Observers are often described as being in the class of predictor–corrector methods. The Luenberger observer of Figure 4-8 is shown in Figure 4-13 in two sections. The model of the plant, $G_{PEst}(s)$, predicts the state, $C_O(s)$, from the power converter output signal, $P_C(s)$. Were the model of the plant a perfect representation of the physical plant and the power conversion signal a perfect representation of the power converter output, the prediction would be perfect. Of course, this is not realistic. The model is only an approximation to the plant. In addition, disturbances, which feed into the plant after the power conversion signal, usually are unknown to the controller. Disturbances are shown as $D(s)$ in Figure 4-13; note that disturbances are present in most control systems. (The effect of disturbances on the Luenberger observer will be discussed in detail in Chapter 6.) All of these effects could cause the prediction of $C_O(s)$ to deviate substantially from $C(s)$. However, the observer compensator corrects most of those deviations.

The corrector section processes $C_O(s)$ with the estimated sensor transfer function, $G_{SEst}(s)$, to derive the observed sensor output, $Y_O(s)$. When the observed sensor output

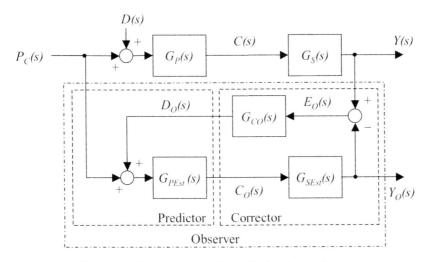

Figure 4-13. Predictor–corrector form of the Luenberger observer.

is compared to the actual sensor output, the observer error, $E_O(s)$, is formed. This error is fed into the observer compensator, $G_{CO}(s)$, to create the signal $D_O(s)$. The gains of $G_{CO}(s)$ are usually set high so that even small errors cause a substantial value of $D_O(s)$. $D_O(s)$ is driven through the plant and sensor models; if the observer is applied properly, the error signal will be held to near zero. Since the correction signal feeds through the plant and the sensor models, it corrects the observed state, $C_O(s)$, as it corrects the observed sensor output, $Y_O(s)$.

4.4 Filter Form of the Luenberger Observer

The Luenberger observer can be analyzed by representing the structure as a transfer function. This method will be used throughout this book to investigate system response to nonideal conditions: disturbances, noise, and model inaccuracy. The form is not normally used for implementation because of practical limitations, which will be discussed. However, the filter form is useful because it provides insight into the operation of observers.

The observer transfer function has two inputs, $P_C(s)$ and $Y(s)$, and one output, $C_O(s)$. In this analysis, the actual model and sensor are considered a black box. This is shown in Figure 4-14. The focus is on understanding the relationship between the inputs to the observer and its output. Signals internal to the observer, such as $E_O(s)$ and $Y_O(s)$, are ignored. In fact, they will become inaccessible as the block diagram is reduced to a single function.

Using Mason's signal flow graphs to build a transfer function from Figure 4-14 produces Equation 4.2. There is a single path, P_1 (see Figure 4-14), from $Y(s)$ to the observed state, $C_O(s)$. Also, there is one path, P_2, from $P_C(s)$ to the $C_O(s)$. Finally, there

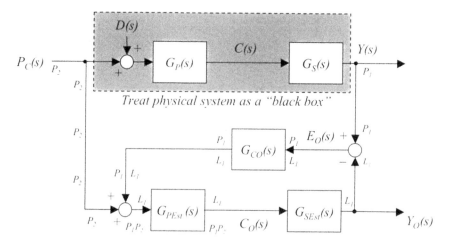

Figure 4-14. Luenberger observer as transfer function between $Y(s)$, $P_C(s)$, and $C_O(s)$.

is a single loop, L_1. The equations for P_1, P_2, and L_1 can be read directly from Figure 4-14:

$$P_1 = G_{CO}(s) \times G_{PEst}(s)$$
$$P_2 = G_{PEst}(s)$$
$$L_1 = -G_{CO}(s) \times G_{PEst}(s) \times G_{SEst}(s).$$

Using Mason's signal flow graphs, Equation 4.2 can be written by inspection.

$$C_O(s) = \frac{Y(s) \times P_1 + P_C(s) \times P_2}{1 - L_1} \qquad (4.2)$$

$$C_O(s) = \frac{Y(s) \times G_{CO}(s) \times G_{PEst}(s) + P_C(s) \times G_{PEst}(s)}{1 + G_{PEst}(s) \times G_{CO}(s) \times G_{SEst}(s)}$$

$$= Y(s) \frac{G_{PEst}(s) \times G_{CO}(s)}{1 + G_{PEst}(s) \times G_{CO}(s) \times G_{SEst}(s)} + P_C(s) \frac{G_{PEst}(s)}{1 + G_{PEst}(s) \times G_{CO}(s) \times G_{SEst}(s)}. \qquad (4.3)$$

Equation 4.2 is the sum of two factors. These factors are separated in Equation 4.3. The first factor, dependent on the sensor output, $Y(s)$, is shown in Equation 4.4; note the term $G_{SEst}(s)$ has been multiplied through the fractional term and divided out of the scaling term, $Y(s)$. Equation 4.4 can be viewed as the sensor output, multiplied by the inverse estimated sensor transfer function, and then filtered by the term on the far right; as will be shown, the term on the right is a low-pass filter. So, Equation 4.4 is the form shown in Figure 4-4 followed by a low-pass filter.

$$Y(s) \times G_{SEst}^{-1}(s) \frac{G_{PEst}(s) \times G_{CO}(s) \times G_{SEst}(s)}{1 + G_{PEst}(s) \times G_{CO}(s) \times G_{SEst}(s)} \qquad (4.4)$$

The second factor of Equation 4.3, dependent on the power converter output, is shown in Equation 4.5. Here the estimated plant transfer function, $G_{PEst}(s)$, has been pulled out to scale the filter term. The scaling term is equivalent to the form shown in Figure 4-5 used to calculate $C_{Est}(s)$. As will also be shown, the term on the right is a high-pass filter. So, equation (4.5) is the form shown in Figure 4-5 followed by a high-pass filter.

$$P_C(s) \times G_{PEst}(s) \frac{1}{1 + G_{PEst}(s) \times G_{CO}(s) \times G_{SEst}(s)} \qquad (4.5)$$

4.4.1 Low-Pass and High-Pass Filtering

The rightmost term in Equation 4.4 can be shown to be a low-pass filter. That term is shown in Equation 4.6. To see this, first consider the individual terms of Equation 4.6. $G_{PEst}(s)$ is a model of the plant. Plants for control systems almost uniformly involve one or more integrals. At high frequencies, the magnitude of this term declines to near zero. $G_{SEst}(s)$ is a model of the sensor. Sensor output nearly always declines at high frequencies because most sensors include low-pass filters, either implicitly or explicitly. The term $G_{CO}(s)$ is a little more difficult to predict. It is constructed so the open-loop gain of the observer ($G_{SEst}(s) \times G_{PEst}(s) \times G_{CO}(s)$) will have sufficient phase margin at the observer crossover frequencies. Like a physical control loop, the compensator has a high enough order of differentiation to avoid 180° of phase shift while the open-loop gain is high. However, the maximum order of the derivative must be low enough so that the gain at high frequency declines to zero. So, evaluating the product of $G_{PEst}(s)$, $G_{SEst}(s)$, and $G_{CO}(s)$ at high frequency yields a small magnitude; by inspection, this will force the magnitude of Equation 4.6 to a low value at high frequency. This is seen because the "1" will dominate the denominator, reducing Equation 4.6 to its numerator.

$$\frac{G_{PEst}(s) \times G_{CO}(s) \times G_{SEst}(s)}{1 + G_{PEst}(s) \times G_{CO}(s) \times G_{SEst}(s)} \tag{4.6}$$

Using similar reasoning, it can be seen that Equation 4.6 will converge to "1" at low frequencies. As has been established, $G_{PEst}(s)$ will usually have one order of integral; at low frequencies, this term will have a large magnitude. $G_{CO}(s)$ will add one order of integration or will at least have a proportional term. Typically, $G_{SEst}(s)$ will be a low-pass filter with a value of unity at low frequencies. Evaluating the product of $G_{PEst}(s)$, $G_{SEst}(s)$, and $G_{CO}(s)$ at low frequency yields a large magnitude; by inspection, this will force the magnitude of Equation 4.6 to 1. (This can be seen because the "1" in the denominator will be insignificant, forcing Equation 4.6 to 1.) These two characteristics, unity gain at low frequency and near-zero gain at high frequency, are indicative of a low-pass filter.

The right-hand term of Equation 4.5 can be investigated in a similar manner. This term, shown in Equation 4.7, has the same denominator as Equation 4.6, but with a unity numerator. At high frequency, the denominator reduces to approximately 1, forcing Equation 4.7 to 1. At low frequencies, the denominator becomes large, forcing the term low. This behavior is indicative of a high-pass filter.

$$\frac{1}{1 + G_{PEst}(s) \times G_{CO}(s) \times G_{SEst}(s)} \tag{4.7}$$

4.4.2 Block Diagram of the Filter Form

The filter form of the Luenberger observer is shown as a block diagram in Figure 4-15. This demonstrates how the observer combines the input from $Y(s)$ and $P_C(s)$. Both of these inputs are used to produce the observed state. The term from $Y(s)$ provides good low-frequency information but is sensitive to noise; thus, it is intuitive that this term should be followed by a low-pass filter. The term from $P_C(s)$

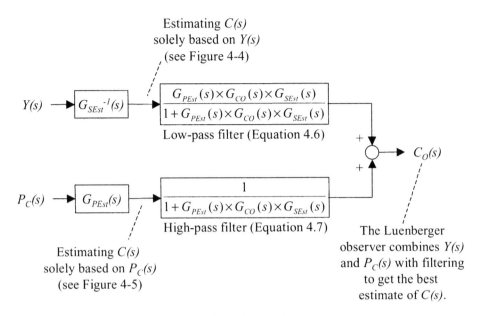

Figure 4-15. Filter form of the Luenberger observer.

provides poor low-frequency information because it is subject to drift when integral gains are even slightly inaccurate. On the other hand, this term is not as prone to generate high-frequency noise as was the $Y(s)$ term. This is because the plant, which normally contains at least one integral term, acts as a filter, reducing the noise content commonly present on the power converter output. It is intuitive to follow such a term with a high-pass filter. The observed state is formed two different ways from the two different sources and uses filtering to combine the best frequency ranges of each into a single output.

One detail that bears discussion is that the two filters of Equations 4.6 and 4.7 sum to unity. This leads to the more computationally efficient version of the filter form that is shown in Figure 4-16. It is based on the understanding that $1 -$ Equation $4.6 =$ Equation 4.7, which can be seen with a little algebra.

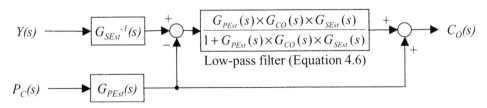

Figure 4-16. Computationally efficient filter form of the Luenberger observer.

4.4.3 Comparing the Loop and Filter Forms

The filter form (Figures 4-15 and 4-16) of the Luenberger observer is presented to aid understanding. The loop form (Figure 4-7) is more practical for several reasons. First, in most cases, it is computationally more efficient. Also, the loop form observes not only the plant state, but also the disturbance; the value of disturbance observation will be discussed in Chapter 6. The filter form does offer at least one advantage over the traditional form: the process of tuning the observer is exchanged for a configuration of a filter; for some designers, this process might be more familiar.

The standard form of the Luenberger observer requires the tuning of the $G_{CO}(s) - G_{PEst}(s) - G_{SEst}(s)$ loop. This process is like tuning a physical control loop. Like a physical control loop, the loop form of observers can become unstable. Assuming the low-pass filter of Figure 4-16 is implemented in the z-domain as a ratio of two polynomials, this type of instability can be avoided entirely. However, this advantage is small as the instability in the loop form is easily avoided. This book will focus on the loop form but use the filter form for analysis.

4.5 Designing a Luenberger Observer

Section 4.2 provided experiments that demonstrated performance of the Luenberger observer. In that section, a fully configured observer was provided as part of a control system. Before the model for that system could be used, several steps were required to design that observer. Those steps are the focus of this section. There are four major components of observer design: modeling the sensor, modeling the plant, selecting the observer compensator, and tuning that compensator. This section will provide details on each of those components and then conclude with a step-by-step procedure.

The presentation of observers in this book is based on classical controls: block diagrams and transfer functions in the s-domain. Readers may have considered that observers are commonly presented in the matrix-based *state-space* form. State space is a useful means of representing observer-based systems, especially when the system is complex, for example, when high-order plants are used. The weakness of the state-space representation is that it hinders intuition. The abstract form simplifies mathematical manipulation but leaves most designers puzzled: How can I implement this? Under what conditions will I see benefit and in what quantity? How do I debug it?

This book presents observers in the classical form, even though the approach is limited to lower order systems. This approach is well used in application-oriented writing concerning observers [15; 33, p. 345; 35; 40; 41], even though it is generally eschewed by authors of academic texts [16, p. 299; 18, p.70; 19, p. 400; 34, p. 653] in favor of the more general state-space form. The classical approach is used here in the belief that a great deal of benefit can be gained from the application of observers to common control systems, but the dominance of the state-space representation has limited its application by working engineers.

4.5.1 Designing the Sensor Estimator

The task of designing the sensor estimator is to derive the transfer function of the sensor. The key parameters are filtering and scaling. For example, the model in Section 4.2 used a low-pass filter with unity gain to model the effects of the sensor. One benefit of modeling the entire control system (as was done in Section 4.2 and throughout this book) is that the transfer function of the sensor is known with complete certainty. In the system of Section 4.2, the sensor was selected as a low-pass filter so that designing the sensor estimator was trivial ($G_{SEst}(s) = G_S(s)$). Of course in a practical system, the sensor transfer function is not known with such precision. The designer must determine the sensor transfer function as part of designing the observer.

4.5.1.1 Sensor Scaling Gain

The scaling gain of a sensor is of primary importance. Errors in sensor scaling will be reflected directly in the observed state. Most sensor manufacturers will provide nominal scaling gains. However, there may be variation in the scaling, either from one unit to another or in a particular unit during operation. (If unit-to-unit variation is excessive, the scaling of each unit can be individually measured.) Fortunately, sensors are usually manufactured with minimum variation in scaling gains so that with modest effort, these devices can often be modeled with accuracy.

Offset, the addition of an erroneous DC value to the sensor output, is another problem in sensors. If the offset is known with precision, it can be added to the sensor model; in that case, the offset will not be reflected in the observed state. Unfortunately, offset commonly varies with operating conditions as well as from one unit to another. Offset typically will not affect the dynamic operation of the observer; however, an uncompensated offset in a sensor output will be reflected as the equivalent DC offset in the observed state. The offset then will normally have the same effect, whether the loop is closed on the sensor output or the observed state: the offset will be transferred to the actual state; in that sense, the offset response of the observer-based system will be the same as that of the traditional system.

4.5.2 Sensor Filtering

The filtering effects in a sensor model may include explicit filters, such as when electrical components are used to attenuate noise. The filtering effects can also be implicit in the sensor structure such as when the thermal inertia of a temperature sensor produces phase lag in sensor output. The source of these effects is normally of limited concern at this stage; here, attention is focused on modeling the effects as accurately as possible, whatever the source. The filtering effects can be thought of more broadly as the dynamic performance: the transfer function of the sensor less the scaling gain. As was the case with scaling gains, most manufacturers will provide nominal dynamic performance in data sheets, perhaps as a Bode plot or as an *s*-domain transfer function. Again, there can be variation between the manufacturer's data and the parts. If the variation is excessive, the designer must evaluate the effects of these variations on system performance.

In some cases, varying parameters can be measured. Since gain and offset terms can be measured at DC, the process to measure these parameters is usually straight-forward. Measuring the dynamic performance of a sensor can be challenging. It requires that the parameter under measurement be driven into the sensor input; the sensor is then calibrated by comparing the sensor output to the output of a calibrating sensing device, which must be faster and more accurate than the sensor. Fortunately, such a procedure is rarely needed. Small deviations between the filtering parameters of the actual and model sensor have minimal effect on the operation of the observer. The evaluation of errors in sensor estimation will be considered in detail in the following chapter. For the purposes of this discussion, the assumption is that sensor model is known with sufficient accuracy.

Since most observers are implemented digitally, filtering effects usually need to be converted from the *s*-domain to the *z*-domain. This was the case in Experiment 4C; note that $G_{PEst}(s)$ and $G_{SEst}(s)$ are digital equivalents to their analog counterparts. The conversion can be accomplished using Table 3-2. This table gives the conversion for one- and two-pole filters; higher order filters can be converted to a product of single- and double-pole filters. The conversion to the *z*-domain is not exact; fortunately, the *z*-domain filters in Table 3-2 provide a slight phase lead compared to their *s*-domain equivalents, so that the digital form can be slightly phase advanced from the analog form. However, when adding the delay introduced by sampling, which affects only the digital form, the result is that both the digital and the analog forms have about the same phase lag below half the sample frequency (the Nyquist frequency).

4.5.3 Designing the Plant Estimator

The task of designing the plant estimator is similar to that of designing the sensor estimator: determine the transfer function of the plant. Also like the sensor, the plant can be thought of as having DC gain and dynamic performance. The plant often provides more challenges than the sensor. Plants are usually not manufactured with the precision of a sensor. The variation of dynamic performance and scaling is often substantial. At the same time, variation in the plant is less of a concern because an appropriately designed observer compensator will remove most of the effects of such variations from the observed state. The following discussion will address the process of estimating the plant in three steps: estimating the scaling gain, the order of integration, and the remaining filtering effects. In other words, the plant will be assumed to have the form

$$G_P(s) = K \times \frac{1}{s^N} \times G_{PF}(s), \tag{4.8}$$

where $G_P(s)$ is the total plant transfer function, K is the scaling gain, N is the order of integration, and $G_{PF}(s)$ is the plant filtering. The task of estimating the plant is to determine these terms.

4.5.3.1 Plant Scaling Gain (K)

The scaling gain, K, is the overall plant gain. As with sensors, the nominal gain of a plant is usually provided by the component manufacturer. For example, a current controller using an inductor as a plant has a gain $1/Ls$ where L is the inductance so that the plant gain, K, is $1/L$; since inductor manufacturers provide inductance on the data sheet, determining the gain here seems simple. However, there is usually more variation in the gain of a plant than in that of a sensor. For example, inductance often is specified only to ±20% between one unit and another. In addition, the gain of the plant during operation often varies considerably over operating conditions. Saturation can cause the incremental inductance of an iron-core inductor to change by a factor of five times or more when current is increased from zero to full current. This magnitude of variation is not common for sensors.

Another factor that makes determination of K difficult is that it may depend on multiple components of the machine or process; the contributions from some of those components may be difficult to measure. For example, in a servo system, the plant gain K is K_T/J, where K_T is the motor torque constant (the torque per amp) and J is total inertia of the motor and the load. The K_T is usually specified to an accuracy of about 10% and the motor inertia is typically known to within a few percentage points; this accuracy is sufficient for most observers. However, the load inertia is sometimes difficult to calculate and even harder to measure. Since it may be many times greater than the motor inertia, the total plant gain may be virtually unknown when the observer is designed. Similarly, in a temperature-controlled liquid bath, the gain K includes the thermal mass of the bath, which may be difficult to calculate and inconvenient to measure; it might also vary considerably during the course of operation.

The problems of determining K, then, fall into two categories: determining nominal gain and accounting for variation. For the cases when nominal gain is difficult to calculate, it can be measured using the observer with modest effort. A process for this will be the subject of Section 4.5.3.4. The problems of in-operation variation are another matter. Normally, if the variation of the plant is great, say, more than 20–50%, the benefits of the observer may be more difficult to realize. Of course, if the variation of the gain is well known, and the conditions that cause variation are measured or can otherwise be determined, the estimated scaling gain can be adjusted to follow the plant. For example, the variation of inductance is repeatable and relatively easy to measure. In addition, the primary cause of the variation, the current in the inductor, is often measured in the control system; in such a case, the variation of the gain can be coded in the observer's estimated plant. However, in other cases, accounting for such variation may be impractical.

4.5.3.2 Order of Integration

Most control-system plants have one order of integration, as is the case with all the plants of Table 2-1. In other cases, the plant may be of higher order because multiple single-order plants are used such as when a capacitor and inductor are combined into a second-order "tank" circuit. Higher order plants are common in complex systems. In any event, the order of the plant should be known before observer design begins. This is a modest requirement for observers since even a traditional control system cannot be designed without knowledge of the order of the plant. The assumption here is that the order of the plant (N in Equation 4.8) is known.

4.5.3.3 Filtering Effects

After the scaling gain and integration have been removed, what remains of the plant is referred to here as filtering effects. The plants considered in this book will normally be simple single- and double-order integrators with scaling. These are the predominant plant types in industry as demonstrated in Table 2-1. Other dynamic effects are presumed to be secondary. As with sensors, filtering effects need to be converted to the z-domain.

Like sensors, the filtering effects of plants can generally be determined through manufacturer's data. In most cases, these effects are small or occur at high enough frequencies that they have little influence on the operation of the control system. For example, servomotors have two filtering effects which are generally ignored: viscous damping and interwinding capacitance. Viscous damping contributes a slight stabilizing effect below a few Hertz. It has so little effect that it is normally ignored (this should not be confused with Coulomb and static friction, both effects that are significant but highly nonlinear around zero speed where they cause serious problems). Parasitic capacitance that connects windings is an important effect but usually has little to do directly with the control system, because the frequency range of the effect is so far above the servo controller bandwidth.

4.5.3.4 Experiment 4E: Determining the Gain Experimentally

As discussed in Section 4.5.3.1, the plant scaling gain often must be determined experimentally. The process to do this is simple and can be executed with an ordinary Luenberger observer assuming the observer error signal (the input to the observer compensator) is available and that there are no significant disturbances to the control loop during this process. The system should be configured as follows:

1. Configure the observer estimated sensor and plant. Use the best guess available for the estimated plant scaling gain.
2. Configure the system to close the control loop on the sensor feedback.

3. Set the observer gains to very low values.
4. Minimize disturbances or operate the product–process in regions where disturbances are not significant. (Disturbances reduce the accuracy of this procedure.)
5. Excite the system with fast changing commands.
6. Monitor $E_O(s)$, the observer error.
7. Adjust the estimated plant scaling gain until the observer error is minimized.

Experiment 4E modifies the model of Experiment 4C for this procedure. The estimated sensor is accurate; the estimated plant less the scaling gain is also configured accurately. Only the scaling gain is in error (20 here, instead of 50). The observer error, $E_O(s)$, has large excursions owing to the erroneous value of K_{Est}. The result is shown in Figure 4-17.

The effect of changing the estimated scaling gain is shown in Figure 4-18. If K_{Est} is either too large (a) or too small (c), the error signal grows. The center (b) shows K_{Est} adjusted correctly ($K_{Est}=50$). Only when K_{Est} is adjusted correctly is E_O minimized.

The process to find K_{Est} is demonstrated here with a near-perfect sensor estimator ($G_S(s)=G_{SEst}(s)$). However, it is still effective with reasonable errors in the estimated sensor filtering effects. The reader is invited to run Experiment 4E and to introduce error in the estimated sensor and to repeat the process. Set the bandwidth of the estimated sensor to 30 Hz, a value 50% high (double-click on the node at the top center of G_{SEst} and set the value to 30). Notice that the correct value of K_{Est} still minimizes error, though not to the near-zero value that was attained with an accurate estimated sensor in Figure 4-18.

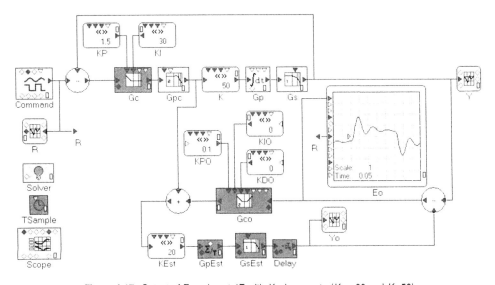

Figure 4-17. Output of Experiment 4E with K_{Est} inaccurate ($K_{Est}=20$ and $K=50$).

Error is minimized
with correct gain.

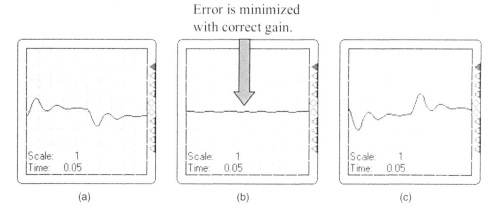

Figure 4-18. Effect of changing K_{Est} on $E_O(s)$ for three settings of K_{Est}: (a) K_{Est}=30; (b) K_{Est}=50; (c) K_{Est}=85.

Notice in Figure 4-18 that the signal is triggered by the rising edge of the command signal, which correlates to the rising edge of plant response. Notice that in Figure 4-18a, where K_{Est} is low, the initial error (synchronized with a positive-going plant) is positive; that is, when the gain is low, the action from the actual plant is greater than that of the observer model. This is intuitive; if the model gain is low, it will react to the excitation less than the actual plant. On the other hand, when K_{Est} is high, as in Figure 4-18c, the gain will be in the opposite direction of plant action indicating that the model reacts more than the actual plant. This can lead to a more automated process where the sign of the error (in comparison to plant action) indicates the direction to adjust K_{Est}.

The procedure that started this section stated that low gains should be placed in the compensator. Experiment 4E used K_{IO}=0, K_{DO}=0, and K_{PO}=0.1, low values indeed. The only purpose here is to drive the DC error to zero. Otherwise, the fully integrating plant would allow a large offset in E_O even if K_{Est} were correct.

Another requirement of compensator gains is that they provide a stable loop. A single integrator in the combined estimated plant and sensor can be stabilized with a proportional gain. This explains why Experiment 4E has K_{PO} as its only nonzero gain. A double integration in that path requires some derivative term. In all cases, the gains should be low; the low gains allow the designer to see the observer error. High gains in a well-tuned compensator drive the observer error to zero too rapidly for this process.

4.5.4 Designing the Observer Compensator

Observer-compensator design is the process of selecting which gains will be used in the compensator; here, this amounts to selecting which combination of P, I, and D gains will be required. The derivative gain is used for stabilization. The order of integration in the G_{PEst}-G_{SEst} path determines the need for a derivative gain in G_{CO}. If the order is two, a derivative gain in G_{CO} will normally be necessary; without it, the fixed 180° phase lag of double integration makes the loop difficult to stabilize. In addition, the phase lag of the G_{PEst}-G_{SEst} path at and around the desired bandwidth of the observer must be considered. If the phase lag is about 180° in that frequency range, a derivative term will normally be required to stabilize the loop.

The need for a derivative term and its dependence on the behavior of G_{PEst}-G_{SEst} is demonstrated by comparing the compensators of Experiments 4C and 4E. In Experiment 4E, the derivative gain is zeroed because the single-integrating plant did not require a derivative term. In Experiment 4C, a derivative term was required. The reason for the difference is the change in observer bandwidth. In Experiment 4E, the observer bandwidth is purposely set very low, well under the sensor bandwidth. In Experiment 4C, the observer is configured to be operational and its bandwidth is well above the 20-Hz bandwidth of the sensor; well above the sensor bandwidth, the single-pole low-pass filter of the sensor has a 90° phase lag like an integrator. That, combined with the 90° phase lag of a single-integrating plant, generates 180° in the G_{PEst}-G_{SEst} path and so must be compensated; the derivative gain of $G_{CO}(s)$ is a simple way to do so.

Beyond the cases where a derivative gain is required in the compensator, derivative gains can be used to enhance stability of the loop (this is generally the case in control systems). However, derivative gains amplify noise. As will be discussed in Chapter 7, observers are particularly sensitive to noise and the derivative gains can needlessly increase that sensitivity.

The goal of the observer compensator is to drive observer error to zero. A fully integrating plant will normally require an integral gain for this purpose. Without the integral gain, disturbances will cause DC errors in the observed state. Because disturbances are present in most control systems and because most applications will not tolerate an unnecessary DC error in the observed state, an integral gain is required in most observer-based systems. If an integral gain is used, a proportional gain is virtually required to stabilize the loop. Thus, most systems will require a PID or, at least, a PI compensator.

The PI–PID compensator is perhaps the simplest compensator available. However, other compensators can be used. The goal is to drive the observer error to zero and to do so with adequate margins of stability. Any method used to stabilize traditional control loops is appropriate for consideration in an observer loop. This book focuses on relatively simple plants and sensors so the PID compensator will be adequate for the task.

One final subject to consider in the design of G_{CO} is saturation. Traditional control-loop compensators must be designed to control the magnitude of the integral when the power converter is heavily saturated. This is because the command commonly changes faster than the plant can respond and large-scale differences in command and response can cause the integrator to grow to very large values, a phenomenon commonly called *wind-up*. If wind-up is not controlled, it can lead to large overshoot when the saturation of the power converter ends. This is not normally a concern with observers. Since they follow the actual plant, they are usually not subjected to the conditions that require wind-up control. For these reasons, PID observer compensators normally do not need wind-up control. When working in *Visual ModelQ* be aware that all of the standard *Visual ModelQ* compensators have wind-up control. Be sure to set the integral saturation levels in the compensator very high so they do not inadvertently interfere with the operation of the observer.

4.6 Introduction to Tuning an Observer Compensator

Tuning an observer compensator is much like tuning a traditional control system. Higher gains give faster response but greater noise susceptibility and, often, lower margins of stability. A major difference is that the observer plant and sensor parameters are set in software; thus, they are known accurately and they do not vary during operation. This allows the designer to tune observers aggressively, setting margins of stability lower than might be acceptable in traditional control loops.

In many cases the gains of the observer compensator are set as high as margins of stability will allow. As observers are almost universally implemented digitally, the main cause of instability in the observer is delay from the sample time of the observer loop. Typically, the observer should operate at a bandwidth much higher than the control loop for which it provides feedback. For this reason, the observer will sometimes need to sample faster than the control loop.

Other issues in observer tuning that must be taken into consideration include noise, disturbances, and model inaccuracy. These are discussed briefly here and will be discussed in detail in the following three chapters. Noise considerations must be considered when tuning an observer. Like any control system, higher gains cause increased noise susceptibility. Further, because of the structure of Luenberger observers, they are often more susceptible to sensor noise than are most control systems. In many cases, sensor noise, not stability concerns, will be the primary upper limit to observer gains.

As with any control system, higher gains provide faster response. The primary advantage to tuning the observer loop with high gains is fast response to disturbances and model inaccuracy, reducing the effects of those problems on the observed state. These subjects will be discussed in detail in Chapters 5 and 6. For the purposes of this discussion, the assumed goal is to tune the observer for maximum gains attainable with adequate margins of stability.

In order to provide intuition to the process of tuning, the Luenberger observer of Figure 4-14 can be drawn as a traditional control loop, as shown in Figure 4-19. Here, the sensor output, $Y(s)$, is seen as the command to the observer loop. The loop is closed on the observed sensor output, $Y_O(s)$; the power converter output, $P_C(s)$, appears in the position of feed-forward term. (Just to be clear, this diagram shows the observer and no other components of the control system.) Accordingly, the process

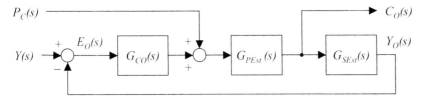

Figure 4-19. Luenberger observer drawn as a traditional control loop.

of tuning the observer loop is similar to tuning a traditional control system with feed-forward.

The procedure here is:

1. Temporarily configure the observer for tuning.
2. Adjust the observer compensator for stability.
3. Restore the observer to the normal Luenberger configuration.

4.6.1 Step 1: Temporarily Configure the Observer for Tuning

The observer should be temporarily configured with a square-wave command in place of Y and without the feed-forward (P_C) path. This is done in Experiment 4F and is shown in Figure 4-20. The command square-wave generator, R, has been connected in the normal position for Y. The estimated sensor output, Y_O, is the primary output for this procedure.

The square-wave generator is used because it excites the observer with a broader range of frequencies than the physical plant and sensor normally can. The rate of

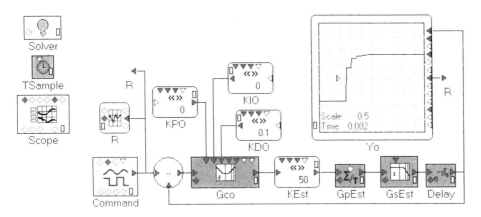

Figure 4-20. Experiment 4F: Tuning an observer compensator.

change of a sensor is limited by physics: inertia makes instantaneous changes un-attainable. There is no such limitation on command generators. The use of the square-wave command to excite the observer allows the designer to more clearly see its response to high frequencies. The sharper edges of a square wave reveal more about the observer's margins of stability than do the gentler output signals of a physical sensor.

4.6.2 Step 2: Adjust the Observer Compensator for Stability

As with physical loops, the observer compensator can be tuned experimentally with a zone-based tuning procedure (see Section 3.4). For a PID observer compensator, this implies that the proportional and integral terms are zeroed while the derivative gain is tuned. The proportional gain is tuned and then, finally, the integral term it tuned. For a PI observer compensator, zero the I-term and tune the P-term; then tune the I-term. The results of applying this method to the PID compensator in Experiment 4F are shown in Figure 4-21. Note that in many cases, a DC offset will appear in the feedback signal when K_{PO} is zero; if the offset is large, it can move the error signal off screen. If this occurs, temporarily set K_{PO} to a low value until the offset is eliminated; in Experiment 4F, use $K_{PO}=0.1$ for this effect.

During the process of tuning, the term driven by P_C should be zeroed. This is because the P_C term acts like a feed-forward to the observer loop, as shown in Figure 4-19. Zone-based tuning recommends that all feed-forward paths be temporarily eliminated because they mask the performance of the loop gains. After the loop is tuned, the path from P_C can be restored.

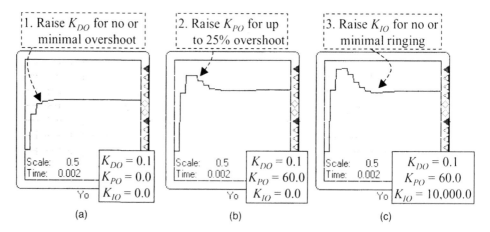

Figure 4-21. Tuning the observer of Experiment 4F in zones. Start with zero gains. (a) Raise K_{DO}, but avoid overshoot. (b) Raise K_{PO} for about 25% overshoot. (c) Raise K_{IO} until ringing just starts.

4.6.2.1 Modifying the Tuning Process for Nonconfigurable Observers

If it is not possible to reconfigure the observer to the form shown in Figure 4-20, the observer can still be tuned by selecting the gain margin. First the system must be configured so the control loop is closed on the sensed signal since this method will temporarily generate instability in the observer. The low-frequency gains are zeroed and the high-frequency gain (typically K_{DO}) is raised until the observer becomes unstable, indicating 0 dB gain margin. The gain is then lowered to attain the desired gain margin. For example, raise K_{DO} in steps of 20% until the observer oscillates, and then reduce it by, say, 12 dB (a factor of 4) to yield 12 dB of gain margin for the observer.

The interested reader may wish to verify this with Experiment 4F. Configure the waveform generator *Command* (right-click on the waveform generator block to adjust its properties) for gentler excitation, similar to what might come from a physical plant and sensor: set Waveform to "s-curve" and set Frequency to 20 Hz. Zero K_{PO} and K_{IO}. (Temporarily set K_{PO} to 0.1 to eliminate DC offset if necessary.) Raise K_{DO} in small steps until the observer becomes unstable — this occurs at $K_{DO} \sim 0.32$. Notice that there are no signs of instability with the s-curve command until the system becomes self-oscillatory. Reduce K_{DO} by 12 dB or a factor of 4 to 0.08 to get 12 dB of gain margin; this is essentially the same value that was found with the tuning procedure above ($K_{DO} = 0.1$). Repeat for the remaining gains.

4.6.2.2 Tuning the Observer Compensator Analytically

Observers can be tuned analytically; this is generally easier than tuning a physical system analytically because the transfer function of the observer is known precisely. After the transfer function is found, the gains can be set to modify the poles of the transfer function. Pole placement methods are presented in [16, p. 308].

4.6.2.3 Frequency Response of Experiment 4G

This section will investigate the frequency response of the observer. Experiment 4G is reconfigured to include a dynamic signal analyzer or DSA (see Figure 4-22). The DSA is placed in the command path; this means the DSA will inject random excitation into the command and measure the response of certain signals. The closed-loop response is shown by the DSA as the relationship of Y_O to R. A variable block has been added for E_O, the observer error, so the open-loop response of the observer can be displayed. The relationship of Y_O to E_O in the DSA is the open-loop gain of the observer.

The closed-loop Bode plot taken from Experiment 4G is shown in Figure 4-23. The plot shows the closed-loop response as having high bandwidth compared to the sample rate. The observer is sampled at 1000 Hz as defined by the digital controller

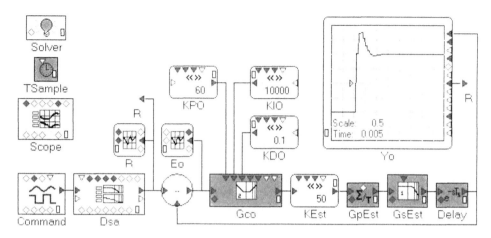

Figure 4-22. Experiment 4G, adding a DSA to Experiment 4F.

(not shown in Figure 4-22). However, the response is still 0 dB at 250 Hz, one-fourth of the sample rate. (The performance above 250 Hz is difficult to determine here since the FFT also samples at 1000 Hz and does not provide reliable data above one-fourth of its sample frequency.) There is about 4 dB of peaking at 100 Hz, a substantial amount for a physical control system but reasonable for an observer. Recall that

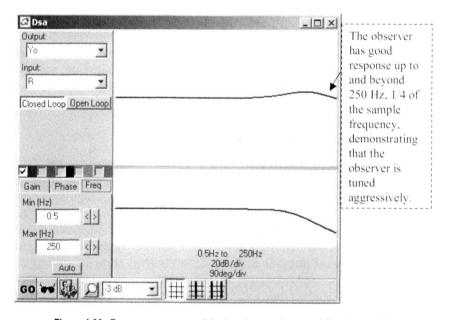

Figure 4-23. Frequency response of the Luenberger observer of Experiment 4G.

observers do not need robustness for dealing with parameter changes as do ordinary control systems and so stability margins can be lower than physical control systems.

Experiment 4G confirms that the observer gains are set quite high for the sample rate. In fact, the only limitation is the sample rate itself. Were the observer sampled faster, the gains could be raised. The reader may wish to experiment here. The sample time (*TSample*) can be reduced to 0.00025. Using the same procedure as above, the gains can be raised to approximately $K_{DO}=0.6$, $K_{PO}=500$, and $K_{IO}=500,000$. In this case the observer bandwidth is about 1000 Hz, still one-fourth of the sample frequency.

One point that needs to be reinforced is that it is often not appropriate to maximize the bandwidth of the observer. High observer bandwidth maximizes the response to sensor noise. (None of the experiments in this chapter have noise sources.) One strength of observers is that they can be used to filter sensor noise while using the power converter signal to make up for any phase lag. So, sensor noise will often be the dominant limitation on observer bandwidth, in which case the observer bandwidth may be intentionally set lower than is possible to achieve based on stability margins. So this procedure reaches the upper limit of observer loop response; be prepared to return and lower the response, for example by lowering K_{DO} and reducing the other gains according to Figure 4-21.

4.6.3 Step 3: Restore the Observer to the Normal Luenberger Configuration

Restore the observer to the normal Luenberger configuration as shown in Figure 4-8. Remove the connection to the waveform generator and reconnect $Y_O(s)$. Reconnect the path from $P_C(s)$. The observer should be ready to operate.

4.7 Exercises

1. Compare the rise time of observer- and nonobserver-based systems to a step command.
   A. Open Experiment 4A and retune the system for reasonable margins of stability (e.g., find maximum K_P without overshoot and maximum K_I for 30% overshoot to step).
   B. What is the settling time?
   C. Repeat A and B using the observer-based system of Experiment 4C.
   D. Why is there a large difference in settling times? How does that difference relate to the observer?
2. Retune an observer for a lower bandwidth.
   A. Open Experiment 4G and retune the system starting with $K_{DO}=0.05$, $K_{PO}=0$, and $K_{IO}=0$. For limits allow 15% overshoot with K_{PO} and slight ringing with K_{IO}.

    B. What is the bandwidth of the observer?

    C. Using Experiment 4C, place the values of part A in the observer and see the effect on command response. Evaluate settling time and compare to results with original observer gains (see Problem 1C). Explain the difference (or lack of difference).

3. Using hot connection on the *Live Scope* in Experiment 4C, compare the plant output (C) to the sensor output (Y).

    A. Are they similar?

    B. If so, does this imply sensor phase lag is not a significant factor in closed-loop performance?

4. Show that having the correct value for estimated sensor bandwidth does not affect the experimental process to find K_{Est} discussed in Section 4.5.3.4. Corrupt the value of the sensor bandwidth by changing it from 20 to 30 Hz (top node of G_{SEst} block); be careful to restore the value to 20 before saving the file as changing the value of a node permanently modifies the *mqd* file.

    A. Adjust K_{Est} to minimize error signal, E_O.

    B. What conclusion could you draw?

Chapter 5

The Luenberger Observer and Model Inaccuracy

I n this chapter . . .

- Common sources of model inaccuracy
- Effects of model inaccuracy on observer-based systems
- Several software experiments demonstrating results of inaccuracy

This chapter continues the discussion of Chapter 4, analyzing and experimenting with the Luenberger observer. Several important aspects of observer performance are presented, especially how observers behave in the presence of errors in the observer model. Key points are developed analytically and demonstrated in software experiments.

In the previous chapter, the observer structure was developed largely assuming ideal conditions. Of course, in practical systems, conditions are not ideal. The control-system designer needs to understand the effects of nonideal conditions on the observer and on the control system as a whole. The main sources of nonideal conditions are model inaccuracy, disturbances, and noise. This chapter will discuss problems of model inaccuracy, Chapter 6 will discuss disturbances, and Chapter 7 will deal with the effects of noise.

5.1 Model Inaccuracy

Model inaccuracy describes the different types of error that can be present in plant and sensor models; these models form the observer's model system as shown in Figure 5-1. Observers produce the observed state, C_O, by driving two signals, the power

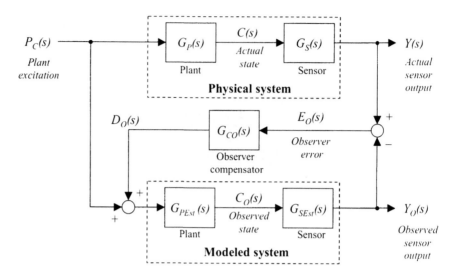

Figure 5-1. The Luenberger observer.

converter output and sensor output, through the plant and sensor models. Of course, the accuracy of the estimated state is directly linked to the accuracy of those models.

5.1.1 Sources of Model Inaccuracy

There are many sources of model inaccuracy. Acquiring data from manufacturers and making independent measurements of control-system components can minimize that inaccuracy. Fine-tuning observer parameters will further reduce inaccuracy. However, in practical systems, some inaccuracy in plant and sensor models will be present. The following sections will address some of the effects of those inaccuracies. The good news, as will be shown in the remainder of this chapter, is that observer-based systems are often no more sensitive to parametric variation than are traditional control systems.

The most common source of model error is probably that the complexity of implementing an accurate model in real time is too great for most systems. Nonlinear effects can be difficult to model in terms of both design effort and requirements for computational resources. For example, friction in motion systems can be modeled as simple viscous damping, which is represented as a force proportional to velocity. A term for Coulomb or *sliding* friction can be added, which is usually represented as a fixed-magnitude force in opposition to the velocity. For greater accuracy at low speeds, the Stribeck effect, which is the cause of stiction, can be modeled. Unfortunately, the Stribeck effect requires considerable computational resources to model [3, 25]. Given the complexity of modeling friction accurately, many motion-system observers rely simply on viscous damping or ignore friction altogether.

Another source of model inaccuracy is the reliance on lumped-parameter models, which reduce model complexity by combining states. For example, the temperature of an object varies across the volume of that object. Modeling only the average temperature lumps the various regions of the object into a single thermal mass, which can be represented by one state. The model's accuracy can be improved by modeling the mass in many discrete regions where each region requires an independent state. However, coarse lumped-parameter models are commonly used because the disadvantage of reduced accuracy is usually offset by the advantages of reduced computational resources and reduced effort to code and configure the observer.

Manufacturers often provide scant detail on the dynamic operation of their products. The responsiveness of a sensor may be given simply as phase lag at a certain frequency or as bandwidth (the $-3\,dB$ frequency), both of which are inadequate representations for an observer model. Observer models require knowledge of component behavior across a broad range of frequencies. Ideally, this would be provided as an s-domain equation or, perhaps, as a Bode plot. However, many manufacturers do not invest the resources to produce such detailed information for their customers.

When characterizing a manufacturer's components, consider contacting the company. Manufacturers may be willing to provide data beyond what they normally publish. In most cases, only a fraction of a manufacturer's customers will require the detailed information needed by someone implementing an observer. If the manufacturer is unable to provide the necessary information on their components, consider making independent measurements. With proper equipment, the dynamic performance of a component can be thoroughly measured. Of course, one disadvantage of making independent measurements is that manufacturers will not stand behind the results. Should the process or materials used to manufacture the components change, the performance of the component may change, possibly rendering the independent measurements inaccurate.

Parametric errors produce errors in the observed state, even when model structures are accurate. This is especially true in complex models because the complexity of measuring or calculating multiple parameters increases with the complexity of the model. Unit-to-unit variation causes parametric inaccuracy. In practical systems, virtually every parameter of a component will vary due to manufacturing tolerances. For example, the torque constant (the relationship between current and torque) of a servomotor will typically vary $\pm10\%$ from one motor to the next. The values of passive electronic components, such as capacitors and inductors, often vary $\pm20\%$. The loss of accuracy in observer models due to unit-to-unit tolerances can be significant.

The variation of sensors is usually more closely controlled than that of plants. Most sensor manufacturers control the variation of key parameters such as DC gain and offset. However, variation in dynamic performance is more difficult to control and generally less important for most control-system applications. As a result, the dynamic behavior of a sensor product line may vary significantly from one unit to another.

One way to compensate for unit-to-unit variation is to fine-tune or "tweak in" the model. However, implementing a process to manually tune every system is usually practical only for low-volume, high-value control systems. It might make sense for a rolling mill, but it probably will not for a power supply. Even when fine-tuning is practical, it adds complications. The process may have to be repeated in the field when components are replaced as part of system maintenance or repair. If the control system is being built for resale, the requirement of field adjustment should be considered carefully as the ability of a company to reliably implement a fine-tuning procedure outside its factory is often limited.

Although the problems of manual fine-tuning make it impractical for many systems, it should be noted that some facets of fine-tuning can be carried out automatically. For example, a method based on zeroing the product of observer error and sensor output $(E_O \times Y)$ can be executed by the controller to fine-tune the scaling constant of the plant model. The use of automated procedures can make fine-tuning practical on high-volume products and for in-field procedures. The $E_O \times Y$ method is demonstrated in Experiment 5B.

Another source of model inaccuracy is component variation during operation. For example, temperature changes the resistance of conductive materials, and the capacitance of electrolytic capacitors varies as the components age. So, the operation of a plant or sensor may change over years of service or just over the time it takes the temperature to change. If these changes are large enough, the accuracy of the observer will be affected, no matter how well the manufacturer's data describe the component or how accurately the model parameters were originally fine-tuned.

While the changes in component behavior with temperature and aging are often small, some operation-dependent system changes are dramatic. For example, the inertia of a reel from which material is unwound may vary by a factor of ten or more over the course of just a few minutes. The liquid level, and thus the thermal mass, of a temperature bath can vary rapidly. These changes can result in variation large enough to cause instability in the control system. In such a case, the condition that causes the change usually must be monitored and the control system adjusted for its effect. Of course, when a parameter varies by an order of magnitude, attempting to characterize it to a few percentage points over the entire operating range is normally impractical.

5.2 Effects of Model Inaccuracy

The effects of model inaccuracy will be evaluated both analytically and experimentally. The strength of analysis is that it allows broad predictions. Unfortunately, analysis is often limited to straightforward effects. The experimental approach offers the ability to investigate a broader combination of effects. The weakness of an experimental approach is that it does not directly yield principles; the ability to predict effects relies on performing experiments under sufficiently broad conditions. The approach in this chapter will be to start with analysis and then experiment with *Visual ModelQ* models.

5.2.1 Analytical Evaluation

The analytical evaluation will be divided into two parts. The first section will discuss the effects of plant-model inaccuracy; the next will look at sensor-model inaccuracy. The focus in this section will be inaccuracy in the scaling gain, here called K_{Est}. This term is often difficult to characterize in the plant model.

5.2.1.1 Plant Inaccuracy

The effect of estimated-plant inaccuracy on the observed state, C_O, can be seen by considering Equation 5.1. (Equation 5.1 was developed in Section 4.4 as Equation 4.3.) The sensor (first) term is followed with a low-pass filter; that filter has a bandwidth equal to that of the observer. Thus, it would be expected that errors in the sensor model would be most significant at frequencies below the observer bandwidth. On the other hand, the power converter (second) term, which relies directly on the plant model, is followed with a high-pass filter. Thus, it would be expected that errors in the plant model would be most significant at frequencies above the observer bandwidth. That is, in fact, the case.

$$C_O = Y(s) \times G_{SEst}^{-1}(s) \frac{G_{PEst}(s) \times G_{CO}(s) \times G_{SEst}(s)}{1 + G_{PEst}(s) \times G_{CO}(s) \times G_{SEst}(s)}$$

$$+ P_C(s) \times G_{PEst}(s) \frac{1}{1 + G_{PEst}(s) \times G_{CO}(s) \times G_{SEst}(s)} \qquad (5.1)$$

5.2.1.2 Corruption of Observed-State Gain Caused by Plant Gain Errors

The power converter (second) term in Equation 5.1 is the product of the power converter output and the plant model, processed by a high-pass filter. Well above the observer bandwidth, the filter gain approaches unity because the denominator term of $G_{PEst}(s) \times G_{CO}(s) \times G_{SEst}(s)$ diminishes; this leaves the power converter term as approximately $P_C(s) \times G_{PEst}(s)$. Here, errors in the plant model, $G_{PEst}(s)$, are translated directly to the observed state. By this reasoning, errors in K_{Est} will translate directly to the output well above the observer bandwidth.

5.2.1.3 Corruption of Observed-State Phase Caused by Plant Gain Errors

A second effect of inaccurate estimated-plant gain is in the phase of the observed state. This effect can also be seen in Equation 5.1. First, consider the actual plant. As discussed in Section 4.1.2, in the absence of inaccuracy, $C(s)$ is equal to both

$P_C(s) \times G_P(s)$ and $Y(s) \times G_S^{-1}(s)$, so that $P_C(s) \times G_P(s) = Y(s) \times G_S^{-1}(s)$. Accordingly, in the ideal observer, $P_C(s) \times G_{PEst}(s) = Y(s) \times G_{SEst}^{-1}(s)$.

Now consider the filtering terms in Equation 5.1. They are low- and high-pass filters with identical bandwidths. Thus, the phase from the filter terms will be equal in magnitude and opposite in sign at all frequencies. In the ideal case, where $P_C(s) \times G_{PEst}(s) = Y(s) \times G_{SEst}^{-1}(s)$, the phase from the two filtering terms will cancel; thus, the phase from the filters will not affect the observed state. However, when K_{Est} is inaccurate, it forces $P_C(s) \times G_{PEst}(s)$ high or low; $P_C(s) \times G_{PEst}(s)$ is no longer equal to $Y(s) \times G_{SEst}^{-1}(s)$, and the phase terms from the two filters no longer cancel. The net effect is that errors in K_{Est} cause phase distortion in the observed state around the observer bandwidth. Scaling errors do not cause phase errors well above the observer bandwidth because the magnitude of the low-pass filter term is so low it has little effect on the observed state; also, K_{Est} does not affect the phase of the high-frequency term.

5.2.2 Effects of Inaccuracy in the Sensor Model on C_O

The effects of sensor-model inaccuracy can also be understood by considering Equation 5.1. The sensor model appears mainly in the sensor (first) term, which is the dominant term at low frequency. Thus, the frequency range where inaccuracy in the sensor model dominates is below the observer bandwidth. Consider from Figure 4-4 that $C(s) = Y(s) \times G_S^{-1}(s)$; from the discussion above, at frequencies below the observer bandwidth, the observed state, $C_O(s) \approx Y(s) \times G_{SEst}^{-1}(s)$. Combining this equation and approximation, the observed state at low frequencies compared to the actual state is $C_O(s)/C(s) \approx G_S(s)/G_{SEst}(s)$. Inaccuracy in $G_{SEst}(s)$, which is in the denominator, causes the inverse effect in the observed state, $C_O(s)$, which is in the numerator. For example, phase lag in the estimated sensor model will cause phase advance in observer output.

▒ 5.3 ▒ Experimental Evaluation

Studying inaccuracy requires differentiating between two separate problems: inaccurate modeling and the effects of variation. The difference is whether the error appears before or after the system is configured. If an observer model is incorrect from the start, the error is present before the observer is tuned. The second type of error, variation, appears after the observer is configured. Here, the observer is tuned with accurate models, but through time, operating parameters, or unit-to-unit variation, the dynamics change. The investigation of variation will rely on varying the actual plant and sensor parameters after the observer is configured. The investigation of inaccurate modeling will be performed by varying observer-model parameters before the observer is configured.

5.3.1 Precise Tuning Procedure

Before experiments can be run, a precise observer-tuning process must be developed. The process that will be used in this section is similar to that developed in Chapter 4, but measures of performance will be evaluated more accurately. This effort is not normally required for observers in products—small differences in tuning parameters have minimal effect on system performance. However, in this case, the goal will be to demonstrate general principles using experimental data. Observer dynamic performance should be held as constant as possible to ensure that the data show effects genuinely caused by the error sources under study and not because of unintentional changes in observer tuning.

The procedure that will be used in these sections is:

1. Configure the observer for tuning, similar to that shown in Figure 4-22 (Experiment 4G). Add the ability to adjust the actual sensor bandwidth. The model for this procedure is Experiment 5A and is shown in Figure 5-2. (Note that this model has an observer sample time of 0.0001 s, much faster than the observers of Chapter 4. The change was made to minimize the effects of the sample time so this and the following experiments can focus on model inaccuracy.)

2. As before, start tuning K_{DO} after zeroing K_{PO} and K_{IO}. Tune K_{DO} for 120-Hz bandwidth. The value of bandwidth is somewhat arbitrary. In a physical system it would often be set based on noise considerations, which will be discussed in Chapter 7. For now, 120 Hz will be assumed as a system requirement and all observer tuning will conform to that measure within a few percentage points. Be aware that the bandwidth is not simply the frequency where the observer gain falls to 3 dB,

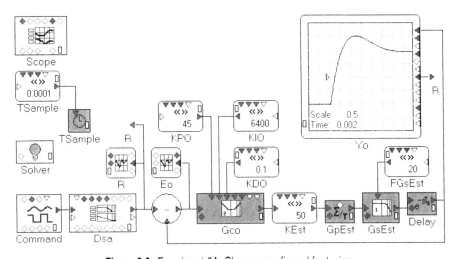

Figure 5-2. Experiment 5A: Observer configured for tuning.

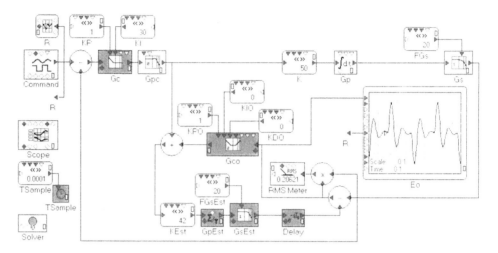

Figure 5-3. Experiment 5B: Adjusting K_{Est} using an RMS meter on $E_O \times Y$.

but where it falls 3 dB below the DC gain. That distinction is important when K_{DO} is adjusted with other gains zeroed because the DC gain is less than zero by a few decibels.

3.   Raise K_{PO} until there is 15% overshoot in response to a step function. As it turns out, this will raise the bandwidth by approximately 30%.

4.   Raise K_{IO} until there is 25% overshoot in response to a step function. This will not substantially affect the bandwidth.

5.   Configure a second model to adjust K_{Est}. The system is similar to Figure 4-17 (Experiment 4E) except an RMS meter has been added to monitor $E_O \times Y$ and so allow more accurate adjustment of K_{Est} through minimizing the meter output. This technique is based on the discussion in Section 4.5.3.4. Using this technique under nominal conditions, K_{Est} will adjust to K with accuracy. The model to adjust K_{Est} is Experiment 5B and is shown in Figure 5-3.

This procedure was applied to the model system of Experiment 5A. The resulting observer gains were $K_{DO}=0.1$, $K_{PO}=45$, and $K_{IO}=6400$, values similar to those derived in Chapter 4. The resulting Bode plots of the observer, less the power converter path, are shown in Figure 5-4.

5.3.2 Simulating Parameter Variation

The effects of plant gain error will be demonstrated by varying the value of K, and the effect of sensor dynamic variation will be simulated by changing the sensor bandwidth. The effects will be considered in two stages. First, observer inaccuracy will be considered. In this case, the control loop will be closed based on sensor feedback. The

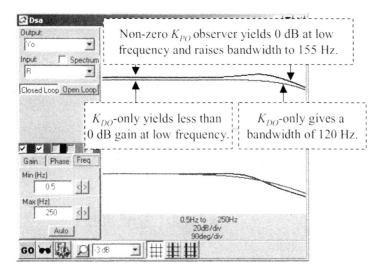

Figure 5-4. From Experiment 5A, gain (above) and phase of frequency response of observer less power converter path for two conditions: K_{DO}=0.1, K_{PO}=K_{IO}=0 and K_{DO}=0.1, K_{PO}=45, K_{IO}=6400.

primary concern will be the comparison of the actual and observed states. Second, the effect on the stability of the overall control loop will be studied. Here, the loop feedback will be taken from the observed state. The primary concern will be the response of the actual state to the command.

5.3.2.1 Effects on the Observed State

The effect of variation on the observed state will be demonstrated using Experiment 5C, which is shown in Figure 5-5. This is a control system with an observer, but still using the sensed signal for feedback; this allows variation in the observed state without affecting control-loop operation. The observed and actual states are shown in *Live Scopes*. *Live Constants* are provided for K (plant gain) and FG_S (sensor bandwidth). While these displays are helpful, time-domain plots of the observed and actual states are not reliable measures of observer performance because many of the effects occur at high frequency. Thus, these experiments will rely mostly on the output of the DSA, which can be brought to view by double-clicking on the DSA icon after the model is compiled. Note also that the gains of the PI control law, G_C, have been reduced (K_P=0.6, K_I=12) so the system will have reasonable margins of stability under nominal conditions. This is done to avoid the distraction caused by the control system ringing.

The effects of varying K are shown in Figure 5-6. Three plots are run for the cases of K=25, 50 (nominal), and 100. The observer accuracy with nominal values is nearly perfect, as is shown with the center plots, which have 0 dB gain and 0° phase lag. When

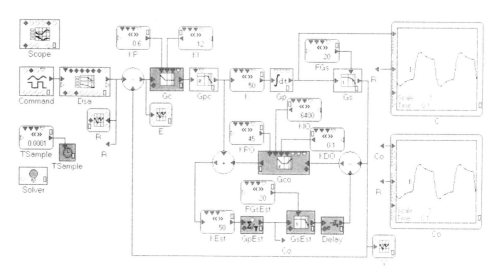

Figure 5-5. Experiment 5C: Investigating the effects of variation on observer performance.

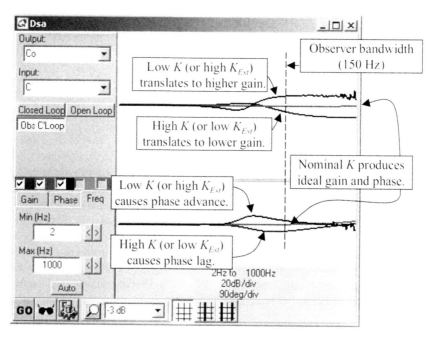

Figure 5-6. From Experiment 5C, the effects on observer accuracy of varying K high (100) and low (25) from nominal (50).

K is raised above K_{Est}, the observer gain is low (that is, K_{Est} is lower than K) and phase lag is injected near the observer bandwidth. When K it lowered, the opposite effects occur. This is consistent with the analytical results above.

The effects of varying the sensor bandwidth, FG_S, on observer accuracy are shown in the Bode plot of Figure 5-7. These results are also from Experiment 5C. Three plots are run for the cases of $FG_S = 10$, 20 (nominal), and 40. When FG_S is nominal, the results are essentially ideal. When FG_S is low (10 Hz), it causes phase lag below the observer bandwidth. This is consistent with the analytical prediction since, below the bandwidth, the $C_O(s)/C(s) \approx G_S(s)/G_{SEst}(s)$ (see Section 5.2.2). Thus, phase lag in $G_S(s)$ should be (and is) translated to phase lag in $C_O(s)$. However, at higher frequencies, the effect of the filtering terms reverses this trend so that low FG_S causes phase advance. Given that lowering FG_S both advances phase and attenuates gain at higher frequencies, one might predict that the overall effect of this error on control-loop stability would be positive. With similar reasoning, raising FG_S should harm control-loop stability. In fact, that is the case when the control-system dynamics are near the observer dynamics. However, that effect will not be demonstrated in these experiments since the observer dynamics are well above the control-loop dynamics.

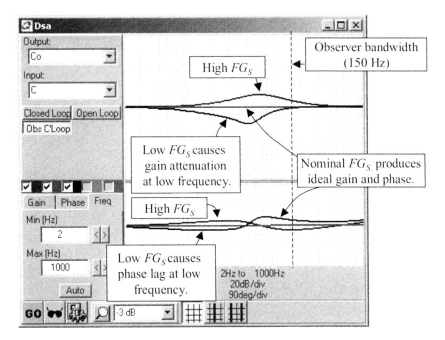

Figure 5-7. From Experiment 5C, the effects on observer accuracy of varying FG_S high (40) and low (10) from nominal (20).

5.3.2.2 Effects on Control-System Response and Stability

The effect of variation on the observed state will be demonstrated using Experiment 5D, which is shown in Figure 5-8. This is similar to Experiment 5C except for two changes. First, the control law (G_C) uses the observed state for feedback. Second, the gains of the PI control law G_C have been raised to $K_P = 1.5$ and $K_I = 30$, the gains used in Chapter 4. The higher control-law gains will make the effects of variation easier to recognize.

The effects on control-loop stability of varying K are shown in Figure 5-9. The bandwidth of the controller varies approximately in proportion to the variation of K. K set to 25, 50 (nominal), and 100 produces about 12-, 25-, and 50-Hz bandwidth, respectively. This variation is neither greater nor less than would be expected in a traditional (nonobserver-based) system. The interested reader can use Experiment 5C to verify this; the bandwidth of that system is 5, 10, and 20 Hz for the cases of $K = 25$, 50, and 100, respectively, which is about 40% of the observer-based system bandwidth in all cases. Also, notice that peaking did not increase substantially for any of the three cases in Figure 5-9. Thus the observer-based system has about 2.5 times the bandwidth of the traditional system and, in this case, is no more sensitive to variation in plant gain.

The effects on control-loop stability of sensor variation are shown in Figure 5-10. System bandwidth remains about constant, but peaking in the closed loop increases considerably, ranging up to 7.5 dB peaking (about 4 dB over the nominal value of 3.2 dB). This is not the result of observer dynamics in and around the observer bandwidth, but rather the simple loss of phase margin at low frequencies as could have been predicted from Figure 5-7.

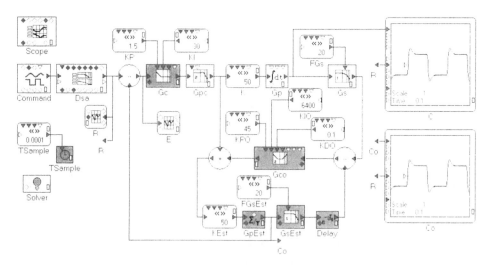

Figure 5-8. Experiment 5D, designed to investigate the effects of variation on control-loop stability.

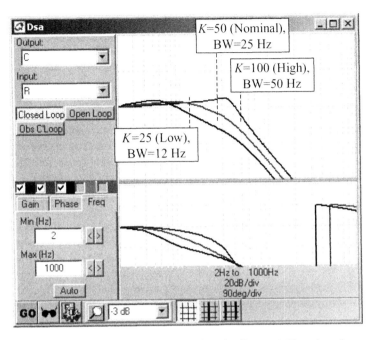

Figure 5-9. From Experiment 5D, effects of varying K on control-loop dynamics.

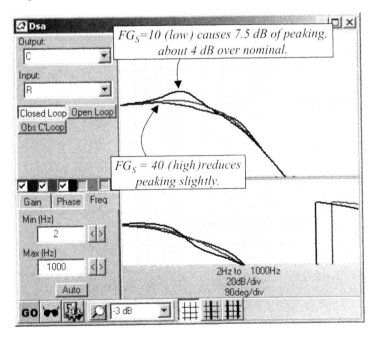

Figure 5-10. From Experiment 5D, effects on control-loop dynamics of setting FG_S to 10, 20 (nominal), and 40 Hz.

5.3.3 Simulating Inaccurate Sensor Modeling

The problem of inaccurate modeling is fundamentally different than that of variation. Variation is a problem of the actual plant and sensor that occurs after the observer has been tuned; inaccurate modeling is a problem of the observer plant and sensor that occurs before the observer is tuned. When studying variation, it is appropriate to vary the actual plant and sensor parameters (such as K and FG_S) while the model is running. When studying inaccurate modeling, it is appropriate to vary the observer parameters. With inaccurate modeling, retuning the observer will mitigate the effects of the inaccuracy. Retuning after introducing error represents the processes that would occur in the application. Failing to retune in the simulation causes the inaccuracy to corrupt the dynamics of the observer loop in a way that would normally not occur and so is unrealistically negative.

Given that the system will be retuned after error is introduced into the observer models, there is little need to investigate the inaccuracy of K_{Est}. The reason is that K_{Est} can usually be determined experimentally at the time of tuning so that modeling inaccuracy is not an issue. On the other hand, when FG_{SEst} does not accurately represent FG_S, problems will result. In the following sections, FG_{SEst} will be set to 10, 20 (nominal), and 40 to investigate the effects of modeling inaccuracy.

5.3.3.1 Tuning with Different Values of FG_{SEst}

The tuning procedure of Section 5.3.1 yields the values shown in Table 5-1. The values of observer-compensation gains vary considerably to adjust for the changes in FG_{SEst}. The key point is that for all values of FG_{SEst}, these gains produce observer dynamics that are about the same. Finally, notice that the setting for K_{Est} is accurate even when the value of FG_{SEst} is off by 2:1. This is further evidence that, for this structure at least, the experimental method can be a reliable way of determining K_{Est}, even in non-ideal conditions.

5.3.3.2 Effects on the Observed State

The effect of the sensor model on the accuracy of the observer is studied using Experiment 5C, which is shown in Figure 5-5. The effects of the changes in FG_{SEst} are shown in Figure 5-11 where FG_{SEst} is adjusted to 10, 20, and 40 Hz. The effects are similar

TABLE 5-1 TUNING VALUES FROM THE PROCEDURE OF SECTION 5.3.1 FOR THREE VALUES OF FG_{SEst}

	$FG_{SEst} = 10$	$FG_{SEst} = 20$	$FG_{SEst} = 40$
K_{DO}	0.2	0.1	0.042
K_{PO}	70	45	35
K_{IO}	9100	6400	3780
K_{Est}	50	50	50

to what was found from changing FG_S except in the opposite directions. Raising FG_{SEst} (and retuning the observer) is approximately the same as lowering FG_S. From the analytical discussion in Section 5.2.2, this is as expected since the key issue is the ratio of $G_S(s)$ and $G_{SEst}(s)$; lowering one has much the same effect as raising the other. Note that, in Figure 5-11, for each value of FG_{SEst}, the tuning gains were adjusted according to Table 5-1.

5.3.3.3 Effects on Control-System Response and Stability

Experiment 5D, as shown in Figure 5-8, is used to study the effect on control-loop stability of erroneous values of FG_{SEst}. The results are shown in Figure 5-12. Again, for each value of FG_{SEst}, the tuning gains were adjusted according to Table 5-1. The effects are similar to changing FG_S (refer to Figure 5-10) except again the directions of the two parameters are opposite. Also, because lowering FG_{Est} causes the gain of the feedback signal to peak at and around 30 Hz (see Figure 5-11), it induced significant peaking in the closed-loop system, forcing reduction of K_P to 1.2. Without this reduction, the change in FG_{Est} with the higher loop gains ($K_P = 1.5$, $K_I = 30$) induced clear signs of marginal stability. The interested reader can confirm by reviewing the

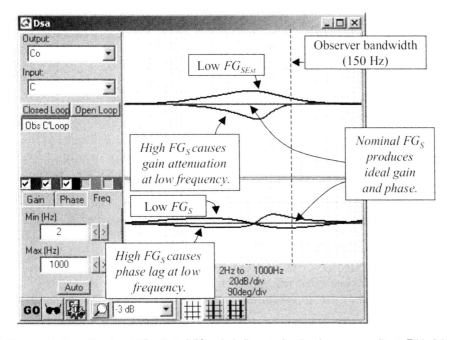

Figure 5-11. From Experiment 5C, effect of FG_{SEst}, including retuning the observer according to Table 5-1.

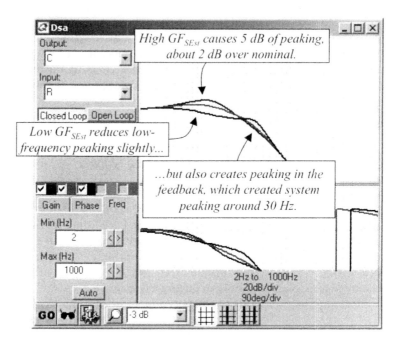

Figure 5-12. From Experiment 5D, effect of varying FG_{SEst}.

step response in Experiment 5D with the following parameters: $FG_{Est} = 10$, $K_P = 1.5$, $K_I = 30$, (from Table 5-1) $K_{DO} = 0.2$, $K_{PO} = 70$, $K_{IO} = 9100$.

5.3.3.4 Detecting Errors in Sensor Dynamics with the Observer

The observer can be used to indicate many parameter errors in the models. Recall that Experiment 5B was used to set the value of K_{Est} based on the RMS value of the term $E_O \times Y$ with the observer gains set low. This same model can be used to evaluate the accuracy of FG_{SEst}. The output of this model is shown in Figure 5-13 where FG_{SEst}, the estimated sensor bandwidth, is 10 Hz (recall that the actual sensor, FG_S, was 20, the nominal value). The error, E_O, is visably larger than when FG_{SEst} was correct, as shown in Figure 5-3. The error here can come from only two sources: K_{Est} and FG_{SEst}. However, it did not come from K_{Est} because that value was adjusted to minimize the RMS meter. Therefore, it must be FG_{SEst}.

In Experiment 5B, FG_{SEst} can be set empirically by adjusting it until the error signal is minimized. (The interested reader can return to Experiment 5B, set FG_{SEst} to 10, and see that the only way to minimize E_O is for both K_{Est} and FG_{SEst} to be set accurately.) This provides a process for configuring observers: Ensure that E_O is nearly zero when observer-compensator gains are low to verify the model is correct. There are

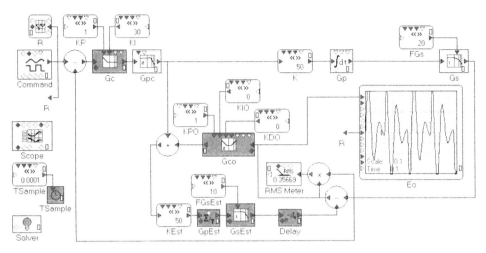

Figure 5-13. From Experiment 5B, the signal E_O clearly indicates model errors. FG_{SEst} is 10 here, but the actual sensor has a bandwidth of 20 Hz; K_{Est} is 50, which is equal to K (compare to Figure 5-3).

errors this process will not detect. It indicates only that $G_P \times G_S \approx G_{PEst} \times G_{SEst}$, as opposed to either $G_P \approx G_{PEst}$ or $G_S \approx G_{SEst}$. Also, it does not provide reliable results in the presence of strong disturbances. Still it can be a useful means of verifying models as it will reveal many types of errors that will be present in practical systems.

5.3.4 Caution About the Experimental Evaluation

The purpose of Section 5.3 was to demonstrate an experimental approach. Experimentation allows designers to evaluate complex effects without mathematics. Designers must be cautious to design experiments that are valid for their machines or processes. Accurate representation of control-system performance and the observer dynamics that will be used on the machine or process are necessary to produce useful results.

This set of experiments has demonstrated at once the strength and weakness of experimental analysis. On the one hand, experiments allow the designer to investigate the effects of nearly any combination of factors. On the other hand, the results may apply only to a narrow set of circumstances. For example, these experiments revealed a system relatively insensitive to variation. However, had the observer bandwidth been closer to the control-system bandwidth, the results would have been different. The designer must be cautious to avoid the overly broad application of experimental results.

5.4 Exercises

1. Tune an observer according to the procedure of Section 5.3.1.
   A. Tune the observer of Experiment 5A for 80-Hz bandwidth.
   B. Repeat for 60-Hz bandwidth.
2. Adjust K_{Est} to match K as well as possible using the RMS meter in Experiment 5B.
   A. Using all the default parameter values, adjust K_{Est}.
   B. Intentionally corrupt estimated sensor bandwidth by changing the actual sensor bandwidth to 25. Repeat A.
3. Describe the change in observer performance for a system with varying sensor dynamics using Experiment 5D.
   A. Execute a Bode plot using *DSA*. Click the DSA button *Obs C'Loop* to display observer closed-loop performance. When the plot is displayed, right-click in the display area of the DSA and click "Save as . . ." and *Red*. Using the *Live Constant FGs*, change the actual sensor frequency to 30 Hz. Run a second Bode plot. Compare the two Bode plots.
   B. Repeat for *FGs*=14 Hz.
   C. Which error in *FGs* will likely induce stability problems in the control loop?
4. Evaluate robustness of observer-based system to traditional system.
   A. Measure the nominal margins of stability for the observer-based control loop of Experiment 5D.
   B. Build a table showing gain crossover frequency, PM, phase crossover frequency, and GM for the following values of K: 20, 50 (nominal), and 100. All other parameters should be at their nominal values.
   C. Using Experiment 5C, build a table similar to that of problem 4B for a traditional control loop.
   D. Compare the two tables and discuss.

Chapter 6

The Luenberger Observer and Disturbances

I n this chapter . . .

- Common sources of disturbance to control systems
- Effects of disturbances on traditional and observer-based systems
- Observed disturbance signals and disturbance decoupling
- Software experiments demonstrating effects of disturbance

This chapter will discuss the sources and effects of system disturbances. Similar to Chapter 5, the presentation will analyze the Luenberger observer using transfer functions and confirm results through experimenting with models. In addition, the principle of disturbance decoupling will be presented. This technique is not unique to observer-based systems; however, it will be interesting for many readers in this context because observers serve the method so well.

6.1 Disturbances

Disturbances enter a system between the power converter and the plant, as shown in Figure 6-1 by the input $D(s)$. Disturbances affect almost every type of control system. In a furnace, heat disturbances from the atmosphere or from neighboring furnaces make temperature more difficult to control. Torque disturbances in a motion system generate velocity and position errors. Load currents can act like a disturbance to a power supply, pulling the output voltage away from the target. In each case, an undesired source of power is added to the power converter output and fed to the plant; the result is that the plant state is disturbed.

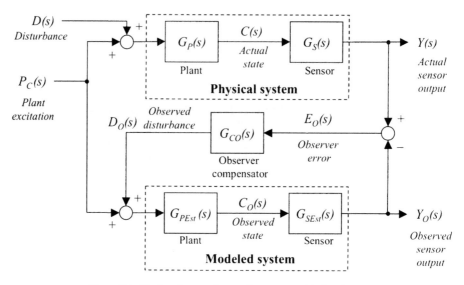

Figure 6-1. The Luenberger observer in a system with disturbances.

For observers, disturbances corrupt the observed state, $C_O(s)$. Recall from Chapter 4 that the command signal fed to an observer through two paths: (1) the command, through a control law, driving the power converter feeds one path of the observer, and (2) the power converter also drives the plant, which then drives the sensor to feed the other path. This combination of prediction and correction gives the observer many of its qualities because the two paths complement each other. Unfortunately, disturbances affect only the sensor (correction) path; they do not benefit from the power converter (prediction) path.

Disturbances cannot be included in the predictor portion of the observer because they are generally unknown. With few exceptions, disturbances are not measured other than indirectly, through their effect on the sensed output. There is normally no way to create the equivalent of a prediction path for them. As a result, disturbances corrupt $C_O(s)$ primarily in frequencies above the bandwidth of the observer, where the correction path of the observer is not effective. Below the observer bandwidth, disturbances have less effect on the accuracy of the observed state because the observer compensator removes their effect.

6.1.1 The Observed Disturbance Signal

The Luenberger observer shown in Figure 6-1 defines the signal exiting the observer compensator as the observed disturbance, $D_O(s)$ [26, 27]. The signal occupies the same position in the observer as the actual disturbance occupies in the actual system, both being added immediately after the power converter output. For the case where the

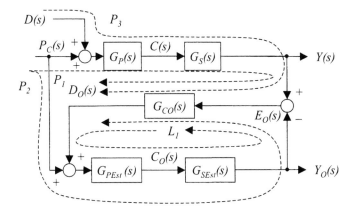

Figure 6-2. Luenberger observer drawn for Mason's signal flow graphs.

plant and sensor models are exact duplicates of their physical counterparts, the only errors in the observed state are those due to disturbances.

The observer compensator must observe the disturbance accurately to keep the observed state accurate. The observer compensator can perform this function well from DC up to the observer bandwidth. At frequencies above the observer bandwidth, the output of $G_{CO}(s)$ no longer follows the actual disturbance well, and the observed disturbance becomes inaccurate; frequency content in the disturbance well above the observer bandwidth translates directly to error in the observed state.

The transfer function from the power converter and actual disturbance to the observed disturbance, $D_O(s)$, can be formed. This construction considers the sensor as being dependent on the disturbance and the power converter; those two signals are the independent inputs in this construction. The observer, shown in Figure 6-2, is redrawn from Figure 6-1 to show the loop and paths for Mason's signal flow graphs.

There is one loop in this system:

$$L_1 = -G_{PEst}(s) \times G_{SEst}(s) \times G_{CO}(s).$$

There are three forward paths, two from the power converter and one from the disturbance:

$$P_1 = P_C(s) \times G_P(s) \times G_S(s) \times G_{CO}(s)$$
$$P_2 = -P_C(s) \times G_{PEst}(s) \times G_{SEst}(s) \times G_{CO}(s)$$
$$P_3 = D(s) \times G_P(s) \times G_S(s) \times G_{CO}(s).$$

All paths "touch" the single loop so the transfer function is $(P_1 + P_2 + P_3)/(1 - L_1)$.

$$D_O(s) = \frac{P_C(s)(G_P(s) \times G_S(s) - G_{PEst}(s) \times G_{SEst}(s))G_{CO}(s) + D(s)G_P(s)G_S(s)G_{CO}(s)}{1 + G_{PEst}(s)G_{SEst}(s)G_{CO}(s)}$$

$$(6.1)$$

For the ideal case where the plant and sensor models are accurate:

$$G_P(s) \cong G_{PEst}(s) \qquad (6.2)$$

$$G_S(s) \cong G_{SEst}(s). \qquad (6.3)$$

Then, Equation 6.1 reduces to:

$$D_O(s) \cong D(s) \frac{G_P(s)G_S(s)G_{CO}(s)}{1+G_P(s)G_S(s)G_{CO}(s)}. \qquad (6.4)$$

The right side of Equation 6.4 can be viewed as the actual disturbance followed by a low-pass filter with a bandwidth equal to the observer bandwidth (refer to Equation 4.6 for discussion on the filter term). So, if the models of the sensor and plant are accurate, the observed disturbance approximates the actual disturbance below the observer-compensator bandwidth.

6.1.1.1 Experiment 6A: Investigating Observed Disturbance

The observed-disturbance signal can be investigated using Experiment 6A, which is shown in Figure 6-3. This system is similar to the system used in Chapters 4 and 5 with a few exceptions:

- The command is set to zero. This experiment is designed to investigate disturbances so that there is little need for a time-varying command signal.
- The waveform generator *Disturbance* has been connected to the disturbance input. This generator is configured to output a square-wave disturbance. This disturbance is used to trigger the scope.
- A DSA is connected through the disturbance input. The Bode plots generated in this experiment will be related to the dynamics of disturbance observation.
- The loop is closed on the actual feedback signal. The gains for the control law ($K_P=0.6$, $K_I=12$) are taken from Section 5.3.2.1.
- The observed disturbance, D_O, is shown in a *Live Scope*.

As with earlier experiments, *Live Constants* are provided to simplify variation of the plant gain and sensor low-pass filter. The observer-compensator gains are taken from Chapter 5, producing an observer bandwidth of 154 Hz.

The *Live Scope* display of Experiment 6A shows the observed disturbance, D_O. The actual disturbance, D, is a step and D_O looks very much like the step response of a low-pass filter. This confirms Equation 6.4. To investigate the subject further, Experiment 6A provides a Bode plot of the observed vs actual disturbance. The DSA can be brought into view by double-clicking on the DSA block; then, click the *GO* button on the lower left of the DSA display. The result, as shown in Figure 6-4,

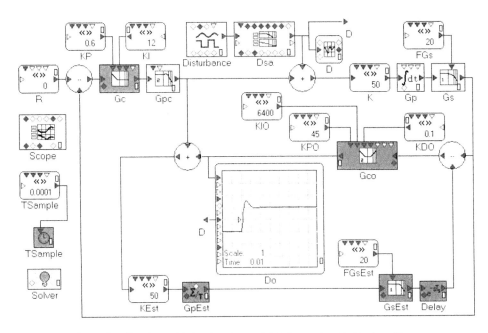

Figure 6-3. Experiment 6A showing observed disturbance signal in ideal conditions.

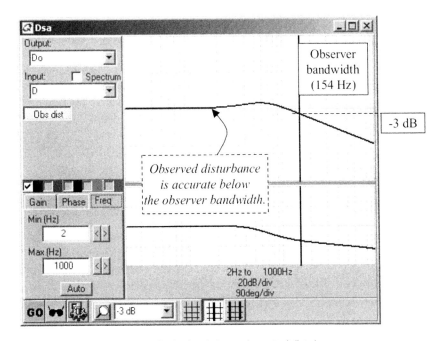

Figure 6-4. Bode plot of observed vs actual disturbance.

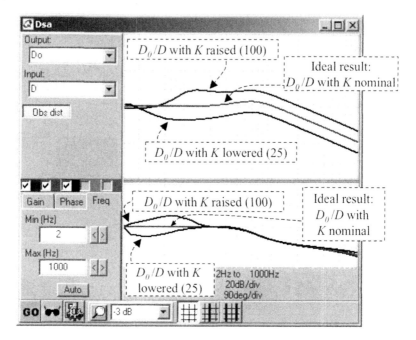

Figure 6-5. Effect of varying K on the accuracy of the observed disturbance.

demonstrates a near-equal (~0 dB) relationship between the two signals up to the observer bandwidth (154 Hz), which is consistent with Equation 6.4.

The close relationship between actual and observed disturbance follows Equation 6.4 only when the plant and sensor models are accurate. Large inaccuracies in the models corrupt the observed disturbance, as is indicated by Equation 6.1. For example, Figure 6-5 shows the effects of K varying from nominal (50) to both high (100) and low (25) values. The effect on the accuracy of the observed disturbance is evident, causing as much as 9 dB (almost 3 times) error in the middle frequencies.

6.1.2 Disturbances and the Integral Term in the Observer Compensator

An integral term in the observer compensator is required to eliminate the effects of steady-state disturbances on the accuracy of the observed-sensor output, $Y_O(s)$. Without the integral term, an offset will appear in the observed state in proportion to the DC disturbance. Most control systems are adversely affected by such an offset. This is why the integral term in the observer compensator is required for most applications.

The effects of DC disturbances can be seen analytically. The goal is to remove DC ($s=0$) error from the observed state ($C_O(s)$) and, thus, from the observed-sensor output

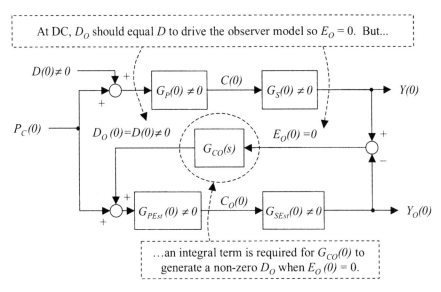

At DC, D_O should equal D to drive the observer model so $E_O = 0$. But...

...an integral term is required for $G_{CO}(0)$ to generate a non-zero D_O when $E_O(0) = 0$.

Figure 6-6. Evaluating the ideal observer at DC with a disturbance shows why an integral term is needed in the observer compensator.

$(Y_O(s))$ in the presence of DC disturbances. DC errors normally generate undesirable offsets in the observer output, which will be transferred to the actual state when the observed state is used as the feedback signal to the main loop. From Figure 6-1, the observed state is:

$$C_O(s) = (P_C(s) + D_O(s))G_{PEst}(s)$$
$$= (P_C(s) + G_{CO}(s) \times E_O(s))G_{PEst}(s) \qquad (6.5)$$

Similarly, the actual state is:

$$C(s) = (P_C(s) + D(s))G_P(s). \qquad (6.6)$$

Assuming that the plant model is accurate so that $G_{PEst}(s) \cong G_P(s)$, the error induced by a disturbance is:

$$C_O(s) - C(s) \cong (G_{CO}(s) \times E_O(s) - D(s))G_P(s). \qquad (6.7)$$

From Equation 6.7, the only way to eliminate all the DC error in the observed state is for $D(0) = D_O(0) = G_{CO}(0) \times E_O(0)$. (This assumes that $G_P(s)$ is nonzero at zero frequency, certainly a reasonable assumption for most practical control systems.) Figure 6-6 shows the observer evaluated at zero frequency ($s=0$). From that figure, if the disturbance, $D(s)$, has a DC component ($D(0) \neq 0$), then in order to eliminate

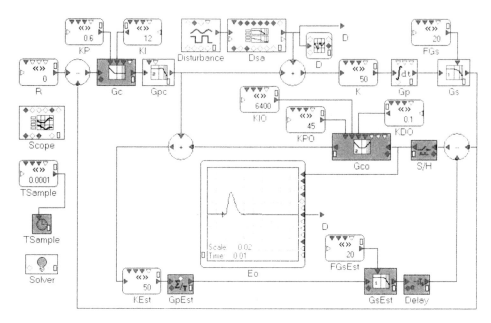

Figure 6-7. Experiment 6B: Investigating the need for integral gain in the observer compensator.

all DC error, $G_{CO}(s) \times E_O(s)$ must have an equal DC component. Thus, in order to remove all DC error from the observed state, $G_{CO}(0) \times E_O(0) \neq 0$; however, removing all error from the observed-sensor output $(E_O(0) = Y_O(0) - Y(0) = 0)$ implies $E_O(0)$ can have no DC component. Thus, $G_{CO}(s)$ must have an integral term; this is the only means for $G_{CO}(0) \times E_O(0) \neq 0$ while $E_O(0) = 0$. This is diagrammed in Figure 6-6.

The argument presented in Figure 6-6 can be extended to the case where the plant and sensor models are not ideal. If there exists scaling error in one or both of those models, the amount of $D_O(s)$ required to cancel a nonzero $D(s)$ will be proportional to the scaling error(s); in any case, $D_O(s)$ will be nonzero if $D(s)$ is nonzero. In the absence of an integral term in $G_{CO}(s)$, any nonzero output of $G_{CO}(0)$ will require a nonzero $E_O(0)$. Again, the integral gain is required to remove DC inaccuracy in the observed state.

Experiment 6B, shown in Figure 6-7, demonstrates the need for an integral term in the observer compensator in the presence of DC error. This experiment is similar to Experiment 6A with a few exceptions. The observer error, E_O, is shown on the design. The scale on this display is small, 0.02 units per division, because the amount of DC error induced by the disturbance is small. The disturbance is a low-frequency square wave; each half-period of the wave is long enough that the DC response of the observer can be seen. The default settings of the observer compensator include an integral gain ($K_{IO} = 6400$). As can be seen in the model, there is no DC component in the observer error signal, E_O. This is the desired behavior.

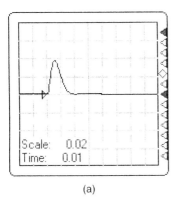 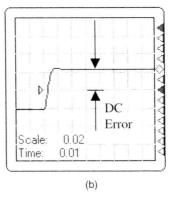

(a)   (b)

Figure 6-8. Results of Experiment 6B, showing that integral gain is required in the observer compensator to eliminate the effects of DC disturbances from the observer error, (a) K_{IO}=6400, DC error is not tolerated in E_O. (b) K_{IO}=0, DC error is tolerated in E_O.

The results of Experiment 6B with nonzero and zero integral gain in the observer compensator are shown in Figure 6-8. In Figure 6-8a, where K_{IO}=6400 (the value used throughout most of the previous chapter), the effect of a disturbance pulse on observer error is transient; after about two divisions (20 ms), the error returns to zero. In Figure 6-8b, the integral gain, K_{IO}, has been zeroed. The observer error remains nonzero indefinitely when in the presence of a DC disturbance. This translates to the equivalent DC error in the observed state in this case because the sensor has unity gain. Were a system implemented using this observed state for feedback, the behavior would be to generate drift when DC disturbances were applied. Such drift, even in small amounts, is normally undesirable.

The reader may have noticed the use of a sample–hold block in Experiment 6B, immediately to the right of the observer compensator. This is a detail more of modeling than of observer operation. It was done to improve the display quality of E_O and has no effect on the operation of the observer. The signal displayed without the sample–hold showed as the comparison of a continuous (Y) and a sampled (Y_O) signal. During the sample period, the continuous signal continued to vary while the sampled signal was fixed. This caused the introduction of a high-frequency component in the error signal. The high frequency aliased down to distort the scope output. The sample–hold here remedied this by synchronizing the error signal to the main digital controller sample time. This has no effect on the operation of the observer because the observer compensator has an implicit sample–hold that performs this same function.

6.2 Disturbance Response

Disturbance response describes to what extent the plant state is perturbed by disturbances. The transfer function of disturbance response is the ratio of the plant perturbation to the disturbance that caused that perturbation: $C(s)/D(s)$. The ideal disturbance response is 0, an unbounded negative number when expressed in decibels. So, control systems are usually structured to minimize the disturbance response. At any given frequency except DC, disturbances will cause some perturbation of the plant state. The goal for most control systems is for that effect to be as small as possible; expressed in decibels, the more negative the disturbance response, the better.

Disturbance response is sometimes described indirectly through the term *stiffness*. Stiffness is the ratio of disturbance to plant perturbation: $D(s)/C(s)$. It is the inverse of disturbance response; accordingly, the higher the stiffness, the better. The terms *disturbance response* and *stiffness* are, of course, equally able to describe the reaction of a system to a disturbance. The former will normally be used in this book.

The primary way observers allow improved disturbance response is by supporting higher control-law gains. As discussed in Chapter 4, the reduced phase lag brought about by using the observed state as the feedback signal increases margins of stability so that control-law gains can be raised. Higher gains improve both command and disturbance response. Disturbance response can also be improved through the use of disturbance decoupling, a technique that is served particularly well by observer-based methods. Disturbance decoupling will be covered in Section 6.3.

6.2.1 Transfer Function of Disturbance Response for Traditional Systems

Disturbance response of a control system both with sensor feedback and with observed-state feedback can be evaluated using transfer functions. The transfer function of the disturbance response of the traditional system shown in Figure 6-9 is easily calculated since there is only one loop.

$$L_1 = -G_C(s) \times G_{PC}(s) \times G_P(s) \times G_S(s)$$

There is a single path from $D(s)$ to $C(s)$: $G_P(s)$. The transfer function is then:

$$\frac{C(s)}{D(s)} = G_P(s)\frac{1}{1+G_C(s) \times G_{PC}(s) \times G_P(s) \times G_S(s)}. \qquad (6.8)$$

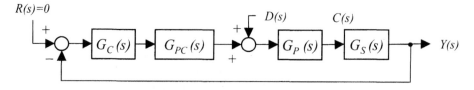

Figure 6-9. Traditional control system.

An algebraic manipulation yields

$$\frac{C(s)}{D(s)} = G_P(s)\left(1 - \frac{G_C(s) \times G_{PC}(s) \times G_P(s) \times G_S(s)}{1 + G_C(s) \times G_{PC}(s) \times G_P(s) \times G_S(s)}\right). \qquad (6.9)$$

Equation 6.9 can be rewritten in terms of the control-law closed-loop transfer function, $G_{CL}(s) = Y(s)/R(s)$:

$$\frac{C(s)}{D(s)} = G_P(s)(1 - G_{CL}(s)). \qquad (6.10)$$

Understanding that the ideal disturbance response is 0, the closer that $G_{CL}(s)$, the closed-loop response, is to unity, the better the disturbance response. The closed-loop response will be closest to one at low frequency and, correspondingly, the disturbance response will be the best. Raising the control-system bandwidth improves disturbance by keeping $G_{CL}(s)$ approximately unity for a wider range of frequencies; the disturbance response of Equation 6.10 will be lower over a wider frequency range, rejecting more of the disturbance input.

Well above the control-loop bandwidth, the closed-loop response will be near zero and the disturbance response will be $G_P(s)$; that is, the disturbances will be limited only by the plant gain. At high frequencies, disturbance response is passive as, for example, when a large capacitor in a power supply prevents high-frequency voltage ripple or when a large inertia prevents high-frequency velocity ripple.

6.2.2 Transfer Function of Disturbance Response when Using Observed-State Feedback

The transfer function of the system with observed-state feedback is similar to Equation 6.10, although the evaluation is more tedious. Here there are three loops, as shown in Figure 6-10:

$$L_1 = -G_C(s) \times G_{PC}(s) \times G_{PEst}(s)$$
$$L_2 = -G_{CO}(s) \times G_{PEst}(s) \times G_{SEst}(s)$$
$$L_3 = -G_C(s) \times G_{PC}(s) \times G_P(s) \times G_S(s) \times G_{CO}(s) \times G_{PEst}(s).$$

All loops touch: L_1 and L_2 through $G_{PEst}(s)$, L_2 and L_3 through $G_{CO}(s)$, and L_1 and L_3 through $G_C(s)$. So,

$$\Delta = 1 - L_1 - L_2 - L_3.$$

There is a single path from the disturbance to the actual state:

$$P_1 = G_P(s).$$

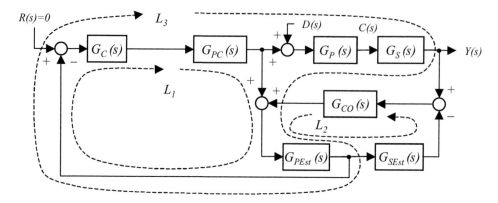

Figure 6-10. System with observed-state feedback.

The path P_1 touches only loop L_3 (through $G_P(s)$), so

$$\Delta_1 = 1 - L_1 - L_2.$$

Assuming accurate observer models ($G_P(s) \approx G_{PEst}(s)$ and $G_S(s) \approx G_{SEst}(s)$), then

$$L_3 \approx -L_1 \times L_2.$$

This leaves:

$$\frac{C(s)}{D(s)} = \frac{G_P(s)(1 - L_1 - L_2)}{(1 - L_1 - L_2 + L_1 L_2)}. \tag{6.11}$$

Two algebraic manipulations provide:

$$\frac{C(s)}{D(s)} = G_P(s)\left(1 - \frac{L_1 \times L_2}{(1 - L_1 - L_2 + L_1 L_2)}\right) \tag{6.12}$$

$$\frac{C(s)}{D(s)} \approx G_P(s)\left(1 - \left(\frac{-L_1}{1 - L_1}\right)\left(\frac{-L_2}{1 - L_2}\right)\right). \tag{6.13}$$

The term $-L_1/(1 - L_1)$ is the closed-loop transfer function of the control system.[1] Similarly, the term $-L_2/(1 - L_2)$ is the closed-loop transfer function of the observer. Substituting these equations into Equation 6.13 provides a result similar in form to Equation 6.10.

---

[1] Note that unlike the traditional system, the closed-loop transfer function of the observer-based system does not include the sensor transfer function. That is because the effect of the sensor is removed when the observer models are accurate.

$$\frac{C(s)}{D(s)} \approx G_P(s)(1 - G_{CL}(s) \times G_{OLPF}(s)) \qquad (6.14)$$

The disturbance response of the observed-state feedback depends on the control-loop transfer function just as it did in the traditional system as shown in Equation 6.10. As discussed in Chapter 4, the observer allows the control gains ($G_C(s)$) to be raised by virtue of reducing phase lag in the loop. This is the primary way in which observers improve disturbance response.

Equation 6.14 also shows that the observer bandwidth can degrade disturbance response. Notice that the form of Equation 6.14 is such that even if the control-loop bandwidth is high (so that $G_C(s)$ remains near unity for a wider frequency span), the observer bandwidth can reduce the disturbance response. This is not a concern if the observer bandwidth is substantially higher than the control-loop bandwidth, as it normally is.

One misconception about observers is that the observer bandwidth can be set arbitrarily low. This technique is sometimes suggested to reduce noise susceptibility (as will be discussed in Chapter 7, noise susceptibility is indeed reduced by reducing observer bandwidth). The reasoning comes from the correct notion that the phase lag caused by the sensor can be removed independently of the observer bandwidth. In that sense, control-law gains can be raised, even if the observer bandwidth is low. In fact, it is possible to lower the observer bandwidth below the control-loop bandwidth. However, the disturbance response is limited by the lower of the observer bandwidth and the control-loop bandwidth, as demonstrated in Equation 6.14. When the control-loop bandwidth is substantially higher than the observer bandwidth, raising control-law gains higher will not benefit the disturbance response. In such cases, higher gains benefit only the command response. This is of questionable benefit in realistic systems, since during the time before the observer loop settles, the command response will only be accurate in the absence of disturbances or model errors.

6.2.3 Improved Disturbance Response Through Control-Law Gains

The primary benefit of Luenberger observers in improving disturbance response is indirect; through the elimination of phase lag from the sensor, the stability margins of the loop are improved. Those improved margins of stability allow increased gains in the main control law, increasing control-system bandwidth. That higher bandwidth improves disturbance response as shown in Equation 6.14.

The benefit of higher control-law gains in improving disturbance response is demonstrated in Experiment 6C, which is shown in Figure 6-11. This model is similar to Experiment 6B, except the actual state is shown in a *Live Scope*. Also, the observed state is used to close the feedback loop. This allows the control-law gains to be raised from (K_P=0.6, K_I=12) to (K_P=1.5, K_I=30) as was discussed in Chapter 4.

The results of Experiment 6C are shown in Figure 6-12. The improvement in disturbance response is evident. The magnitude of the perturbation is increased about

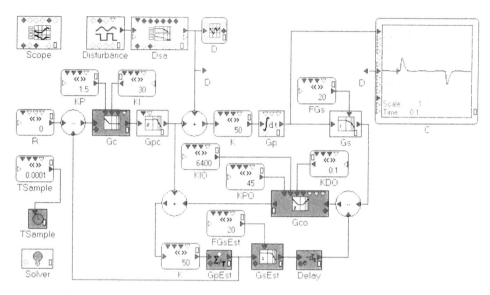

Figure 6-11. Experiment 6C: Investigating disturbance response and control-law gains.

2.5 times when the gains are reduced by that same factor. In addition, the recovery time increases from 0.1 s (1 division) by a factor of, again, 2.5 when the lower gains are used. This behavior is not specific to observers. Most techniques that allow for a substantial increase of control-law gains without compromising stability margins would produce the equivalent enhancement in disturbance response.

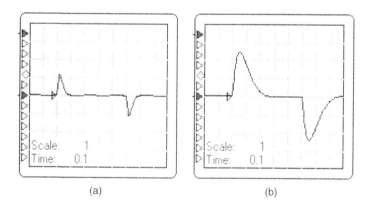

(a)   (b)

Figure 6-12. Results of Experiment 6C: (a) the observer supports high gains ($K_P = 1.5$, $K_I = 30$) that provide superior disturbance response compared to (b) the lower gains of ($K_P = 0.6$, $K_I = 12$) the tradition system.

6.3 Disturbance Decoupling

Disturbance decoupling [11, 26, 27, 39] is a method of improving disturbance response. As discussed above, raising control-law gains is effective at improving disturbance response. However, those gains can be raised only so high because of stability constraints. With disturbance decoupling, disturbance response can often be improved beyond what is possible by raising loop gains.

Disturbance decoupling starts by determining the approximate disturbance. The disturbance signals can be derived from measurement or from model-based methods such as an observer. The disturbance is then *decoupled* from the system by subtracting from the power converter command an amount equal to the disturbance. The disturbance is then reacted to as quickly as the measurement (or observation) allows, within the ability of the power converter. With decoupling, the system reaction to disturbances can be considerably faster than relying wholly on the control law.

Figure 6-13 shows the general form of disturbance decoupling. The actual disturbance is summed with the output of the power converter. Simultaneously, the disturbance is measured or observed. In practical systems, the disturbance can be measured only imperfectly. The imperfection is represented in Figure 6-13 as being measured to a specified accuracy and with a limited bandwidth, as indicated by $G_D(s)$. (For an ideal measurement, $G_D(s)=1$.) In traditional (nonobserver) systems, the accuracy and speed of the measurement depends on the quality of the sensor. The disturbance-decoupling path is scaled by a gain, K_{DD}. This allows, among other things, the decoupling to be turned on ($K_{DD}=1$) and off ($K_{DD}=0$).

Direct measurement of disturbances is impractical for most control systems because of the increased cost and reduced reliability brought about by the addition of a sensor. As an alternative to direct measurement, disturbances can be observed [20, 23, 26, 32, 38] as shown in Figure 6-14. The observed signal, $D_O(s)$, can be accurate and fast if the observer bandwidth is high and the sensor and plant models are accurate, as indicated by Equations 6.2–6.4.

If the observer models are accurate, the observer provides the observed disturbance as a filtered version of the actual disturbance, $D_O(s)=D(s)\times G_{DLPF}(s)$, where $G_{DLPF}(s)$ is defined as the filtering term in Equation 6.4. In other words, the transfer function of $T_{OBSERVER}(s)$ in Figure 6-14 is assumed to simply be $G_{DLPF}(s)$.

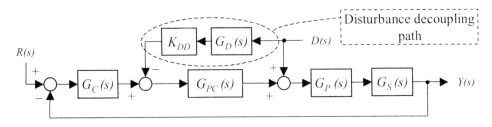

Figure 6-13. Disturbance decoupling.

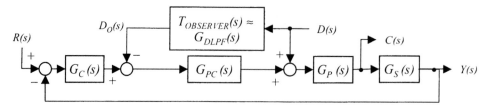

Figure 6-14. Observer-based disturbance decoupling, assuming $K_{DD}=1$.

The transfer function of the disturbance response of the system of Figure 6-14 can be derived using Mason's signal flow graphs and is shown in Equation 6.15. Upon inspection of Equation 6.15, if the disturbance measurement and power converter are ideal (unity), the response to the disturbance would be perfect: zero. This can be seen by assuming $G_{DLPF}(s)$ and $G_{PC}(s)$ were both unity and noticing that the numerator becomes zero, the ideal response to a disturbance.

$$T_{DIST}(s) = \frac{C(s)}{D(s)} = \frac{(1 - G_{DLPF}(s)G_{PC}(s))G_P(s)}{1 + G_C(s)G_{PC}(s)G_P(s)G_S(s)} \qquad (6.15)$$

Of course, the assumption of an unlimited bandwidth observer and power converter is unrealistic. However, Equation 6.15 does demonstrate that for the frequency range below both the power converter and the observer bandwidths, where those transfer functions are approximately unity, disturbance decoupling provides near ideal disturbance response.

6.3.1 Experiment 6D: A Disturbance-Decoupled System

Experiment 6D, shown in Figure 6-15, is a disturbance-decoupled system. This model is almost identical to Experiment 6C, the sole exception being the subtraction of the K_{DD}-scaled observed disturbance, D_O, from the power converter input. Note that the two *Visual ModelQ* extenders named D_O are used to connect the observed disturbance from the observer compensator to the scaling gain, K_{DD}, avoiding confusion that might be caused by crossing lines in the block diagram.

The results of Experiment 6D, as shown in Figure 6-16, demonstrate the improvement available from disturbance decoupling. Figure 6-16a shows the system with K_{DD} set to zero to disable disturbance decoupling; Figure 6-16b shows the results with disturbances decoupled, which is done by setting K_{DD} to unity and zeroing the control-law integrator (the zeroing of the integral will be discussed shortly). The improvement in disturbance response is dramatic: the maximum excursion is cut by more than a factor of two, and the duration of the perturbation is cut by a factor of four. Note that both figures depict the system with the same control-law proportional gain, $K_P=1.5$.

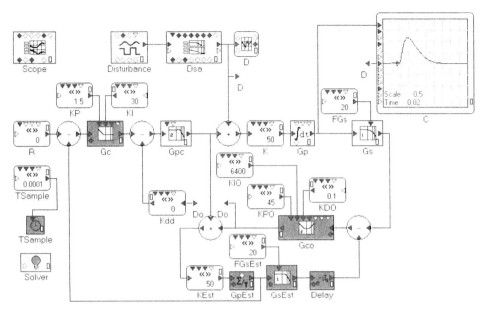

Figure 6-15. Experiment 6D: An observer-based disturbance-decoupled system.

The disturbance response with and without decoupling can be compared using Bode plots. Using Experiment 6D, the Bode plots of disturbance response ($C(s)/D(s)$) are shown with and without disturbance decoupling (see Figure 6-17). At low frequencies, the improvement is 20 dB or a factor of 10. Such an improvement could not be achieved by simply increasing the control-law gains.

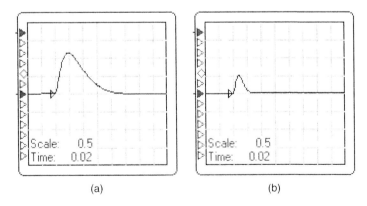

(a)   (b)

Figure 6-16. From Experiment 6D: Disturbance response (a) with a traditional control law ($K_{DD}=0$, $K_I=30$, $K_p=1.5$) and (b) with disturbance decoupling ($K_{DD}=1$, $K_I=0$, $K_p=1.5$).

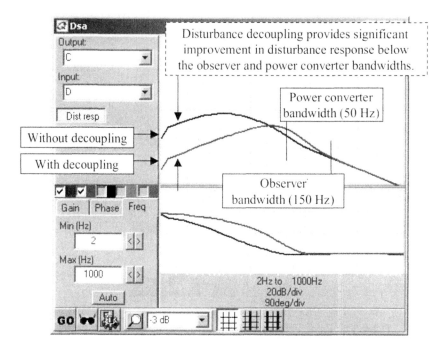

Figure 6-17. From Experiment 6D: Bode plot showing dramatic improvement in disturbance response offered by disturbance decoupling. Plots for two cases: without ($K_{DD}=0$, $K_i=30$) and with ($K_{DD}=1$, $K_i=0$) decoupling; $K_P=1.5$ in both cases.

6.3.2 Disturbance Decoupling Removes Need for Control-Law Integrator

The control law that produced Figure 6-16b is shown in Figure 6-18. Notice that the control-law integrator has been removed (the control law is simply K_P). With disturbance decoupling, the control-law integrator is no longer necessary to provide complete rejection of DC disturbances. This can be seen in the transfer function from $R(s)$ to $C(s)$, which can be derived using Mason's signal flow graphs.

The development that will be performed in this section is somewhat tedious, even when using Mason's signal flow graphs (it can be much more tedious when using competing methods). Readers are encouraged to follow this development as it demonstrates one way to derive and manipulate transfer functions from the block diagrams typical of observer-based systems.

Recall from Section 3.1.4.1 the procedure for Mason's signal flow graphs:

6.3.2.1 Step 1: Find the Loops

The first step is to locate the loops. Figure 6-19 shows Figure 6-18 marked with three loops: L_1, L_2, and L_4. There are two other loops which are not shown in Figure 6-19: L_3 and L_5. Loop L_3 is a figure-eight curve that flows through the power converter,

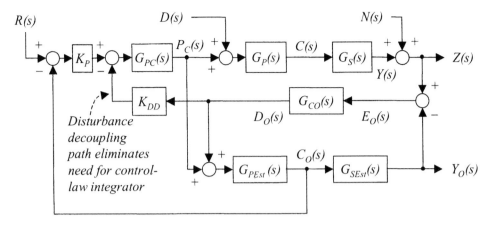

Figure 6-18. When disturbance decoupling is used ($K_{DD}=1$), the control-law integral is no longer necessary.

drops to flow through the observer loop, and then returns through K_{DD}. Loop L_5 starts at the control law, flows through the power converter, plant, and sensor, drops into the observer compensator, then flows through the estimated plant, and returns to the control law. Note that all loops are negative except loop L_3, which is positive since it passes through two subtractions.

The five loops are:

L_1 $-G_{PEst}(s) \times G_{SEst}(s) \times G_{CO}(s)$

L_2 $-G_{PC}(s) \times G_P(s) \times G_S(s) \times G_{CO}(s) \times K_{DD}$

L_3 $+G_{PC}(s) \times G_{PEst}(s) \times G_{SEst}(s) \times G_{CO}(s) \times K_{DD}$

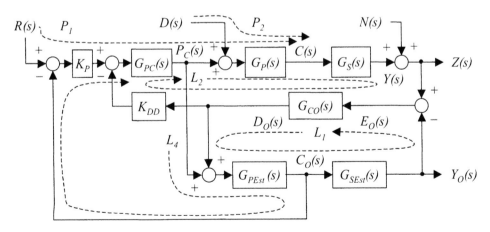

Figure 6-19. Figure 6-18 marked for Mason's signal flow graphs.

$$L_4 \quad -K_P \times G_{PC}(s) \times G_{PEst}(s)$$
$$L_5 \quad -K_P \times G_{PC}(s) \times G_P(s) \times G_S(s) \times G_{CO}(s) \times G_{PEst}(s).$$

6.3.2.2 Step 2: Find the Determinant of the Control Loop

All loops share at least one node. L_1 shares at least one node with L_2, L_3, and L_5 because all pass through $G_{CO}(s)$; L_1 shares at least one node with L_4 because both share $G_{PEst}(s)$. Since L_2 through L_5 all share $G_{PC}(s)$, they must all share at least one node. Thus, the determinant is made of only single loops:

$$\Delta = 1 - L_1 - L_2 - L_3 - L_4 - L_5.$$

6.3.2.3 Step 3: Find All the Forward Paths

There are two paths of concern: P_1 from the command, $R(s)$, and P_2 from the disturbance, $D(s)$:

$$P_1 = R(s) \times K_P \times G_{PC}(s) \times G_P(s)$$
$$P_2 = D(s) \times G_P(s).$$

6.3.2.4 Step 4: Find the Cofactors for Each of the Forward Paths

The cofactor for each forward path is the determinant less the loops that share at least one node with that path. P_1 flows through $G_{PC}(s)$ and so shares at least one node with all loops except L_1. P_2 flows through $G_P(s)$ and so shares at least one node with loops L_2 and L_5.

$$\Delta_1 = 1 - L_1$$
$$\Delta_2 = 1 - L_1 - L_3 - L_4$$

6.3.2.5 Step 5: Build the Transfer Function

The transfer function is then:

$$C(s) = \frac{P_1 \times \Delta_1 + P_2 \times \Delta_2}{\Delta}.$$

A few assumptions simplify the transfer function considerably. First consider that the need for an integrator is best demonstrated at zero frequency. A control-law integrator would not normally be needed if a proportional control law forced $C(s)$ to follow $R(s)$ perfectly at zero frequency, even in the presence of a DC disturbance. At DC, it can be assumed that the power converter is nearly ideal ($G_{PC}(s) \approx 1$) as are the sensor and sensor model ($G_S(s) \approx G_{SEst}(s) \approx 1$). Finally, assume the system is configured for disturbance decoupling, so that $K_{DD} = 1$. So, at zero frequency, L_1 cancels L_3 in Δ and Δ_2:

$$\Delta = 1 - L_2 - L_4 - L_5$$

$$\Delta_2 = 1 - L_4.$$

Filling in the terms for $C(s)$ from above,

$$C(s)|_{s \to 0} = \frac{R(s) \times K_P \times G_P(s) \times (1 + G_{CO}(s)G_{PEst}(s)) + D(s) \times G_P(s) \times (1 + K_P G_{PEst}(s))}{1 + G_P(s)G_{CO}(s) + K_P G_{PEst}(s) + K_P G_{PEst}(s)G_{CO}(s)G_P(s)}.$$

$$(6.16)$$

Now, assume the observer compensator has an integral term. Also, assume the plant and its model are either integrating or nearly integrating, as is the most common case in control systems. So, the functions $G_{CO}(s)$, $G_P(s)$, and $G_{PEst}(s)$ all become very large as s approaches zero. Divide every term in Equation 6.16 by $G_{CO}(s) \times G_P(s) \times G_{PEst}(s)$. Now, to approximate the behavior of Equation 6.16 at $s = 0$, remove every term in Equation 6.16 that has one or more of these three functions in the denominator since those terms will become vanishingly small as s approaches 0. The disturbance term will be cancelled in this process. This yields $C(s) = R(s)$ at DC, the ideal result, showing that an integral in the control law is not necessary.

If the process is repeated for non-integrating plants, those systems where neither $G_P(s)$ nor $G_{PEst}(s)$ becomes large without bound as s approaches 0, it leaves Equation 6.17 at zero frequency:

$$C(s)|_{s \to 0} = R(s)\frac{K_P \times G_{PEst}(s)}{1 + K_P G_{PEst}(s)}.$$

$$(6.17)$$

So, at DC, the system follows the command, independent of the disturbance and very accurately as long as $K_P \times G_{PEst}(0) \gg 1$. If $G_{PEst}(s)$ is a fully integrating model, the DC response of $C(s)$ will be exactly $C(s)$. Notice that there is no requirement that the plant model be particularly accurate as is evidenced by the fact that nowhere in this development was it required for $G_P(s) \approx G_{PEst}(s)$; in fact, the actual plant itself need not even be a true integrator to attain total DC accuracy, so long as the model plant is an integrator. In many cases, plants are not true integrators at very low frequencies because of losses. For example, capacitors have small leakage terms and temperature baths have thermal losses. Both cases are examples of plants that integrate only above some low frequency. However, an integrating model can often be used to simulate such a plant. As long as an integrating model provides a reasonable representation of the plant, it can be used with disturbance decoupling to eliminate the need for an integrator in the control law.

The key step to take note of was where L_1 cancelled L_3 in Δ_2. Without this cancellation, the disturbance term in Equation 6.16 would have a $G_{CO}(s)$ term so it would not have been overwhelmed by the command term at zero frequency.

The reader is encouraged to return to the development and disable disturbance decoupling by setting $K_{DD}=0$. In this case, the DC response for the nondecoupled system without an integral in the control law is:

$$C(s)|_{s \to 0} = R(s) + \frac{D(s)}{K_P}. \qquad (6.18)$$

Without disturbances, the system will follow a DC command perfectly, even without an integrator in the control law. However, the disturbance corrupts the output as is shown in Equation 6.18. A larger proportional gain reduces, but does not eliminate, the error. The most common way to remove all error in the nondisturbance-decoupled system is for the proportional control law, K_P in Equation 6.18, to be replaced with an integrating control law, K_P+K_I/s, as shown in Equation 6.19. The magnitude of such a control law will grow without bound at zero frequency, forcing the disturbance term to vanish from the output. This is the primary reason integrating control laws are so popular in traditional control systems.

$$C(s)|_{s \to 0} = R(s) + \frac{D(s)}{K_P + K_I/s} = R(s) \qquad (6.19)$$

6.3.3 Dynamic Improvement of Disturbance Response

A system employing disturbance decoupling rejects disturbances largely independent of the control law. This is because the primary path to reject disturbances, which is through K_{DD}, does not pass through the control law; it proceeds directly from the observer to the power converter. Here, the bandwidths of the observer and power converters are the primary dynamic limit to disturbance response, as indicated in Equation 6.15. In traditional control systems, the control law is the primary source of disturbance rejection and the control-loop bandwidth, which is usually much lower than the observer or power converter bandwidths.

Experiment 6E, shown in Figure 6-20, will be used to demonstrate how the power converter bandwidth limits disturbance response in disturbance-decoupled systems. This experiment is similar to Experiment 6D, except the bandwidth of the power converter (G_{PC}) is adjusted with the *Live Constant* F_{PC}, both of which are above and left of the center of the figure.

The results of Experiment 6E are shown in Figure 6-21. When the power converter bandwidth is raised from 50 (Figure 6-21a) to 100 Hz (Figure 6-21b), disturbance response improves dramatically. This is consistent with Equation 6.15. Note that the

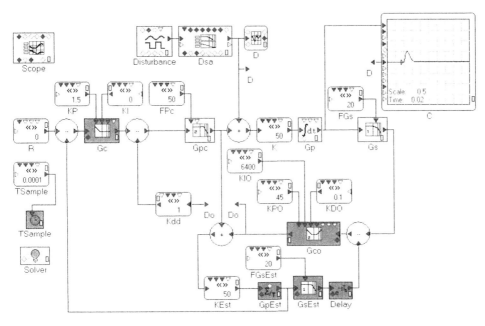

Figure 6-20. Experiment 6E: Investigate the effect of power converter frequency on disturbance response in a disturbance-decoupled system.

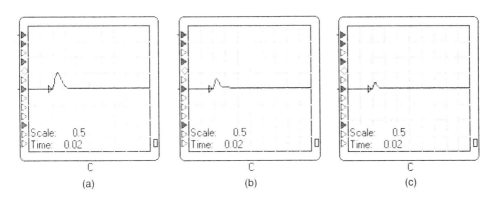

Figure 6-21. From Experiment 6E: Disturbance response with observer-based disturbance decoupling using (a) low (50 Hz), (b) high (100 Hz), and (c) very high (200 Hz) bandwidth power converters.

observer bandwidth is about 160 Hz (from Section 5.3.1), so the power converter, in both cases, is the lower of the two and thus the primary barrier. Accordingly, raising this limit improves disturbance response significantly. Note that in the case where disturbance decoupling is not used, increasing the power converter bandwidth does not directly improve disturbance response to a significant extent. (It can reduce phase lag in the loop, thus allowing higher servo gains, which do improve disturbance response.) The interested reader is invited to confirm this with Experiment 6E using these steps:

- Run Experiment 6E
- Turn disturbance decoupling off (set $K_{DD}=0$, $K_I=30$)
- Adjust F_{PC} up from 50 Hz and observe that disturbance response does not improve significantly. Adjusting it down will affect response because the additional phase lag will cause instability in the control loop.

When the power converter bandwidth is raised further, the disturbance response improves again. The power converter bandwidth of 200 Hz (Figure 6-21c) provides the best decoupling of the three cases in Figure 6-21. However, the benefits of raising the bandwidth are diminishing because the observer bandwidth, which is about 160 Hz, is now the primary limitation on disturbance response. Accordingly, raising the power converter bandwidth from 200 to 500 Hz produces almost no benefit, as the reader will observe by executing Experiment 6E. At this point, the observer bandwidth is the primary limit in Equation 6.15 and substantial improvement in disturbance response can only be achieved by raising that barrier.

6.4 Exercises

1. Compare bandwidth of observed disturbance (D_O/D) and bandwidth of observer. Measure the bandwidth of the observer using Experiment 5A. Use default values for all parameters.
   A. What is the bandwidth of the observed disturbance? (*Hint*: Use the DSA in Experiment 6A.)
   B. What is the bandwidth of the observer itself? (*Hint*: Use the DSA in Experiment 5A.)
   C. Repeat A and B for the observer tuning values $K_{DO}=0.06$, $K_{PO}=25$, and $K_{IO}=1400$.
   D. Repeat A and B for the observer tuning values $K_{DO}=0.04$, $K_{PO}=14$, and $K_{IO}=700$.
   E. Compare the observer bandwidth to the observed-disturbance bandwidth. Discuss the implications of Equations 6.4 and 4.6.
   F. Repeat problem 1D with K set to 20 (leave K_{Est} set to 50). Compare and comment.

2. Find the relationship between the DC error of disturbance in the absence of an integral term in the compensator and the gain K_{PO}. Use Experiment 6B. Set $K_{IO}=0.0$. Note that while the *Live Scope* gives approximate readings, a more accurate measure of error can be obtained with the main scope display, which will come into view after compiling the model, double-clicking on the main scope block (left), and turning on a single cursor (bottom of scope).
   A. Measure the error with the default value of K_{PO} (45).
   B. Repeat with $K_{PO}=20$, 50, and 100.
   C. What is the relationship between the amount of error and K_{PO}?

3. Evaluate disturbance response in the frequency and time domains for three cases: the low gains of the traditional system, the high gains of the observer-based system, and the disturbance-decoupled system. Use Experiment 6D.
   A. Run a Bode plot for the traditional system ($K_P=0.6$, $K_I=12$, and $K_{DD}=0.0$). Save this plot as black. Repeat for the higher gains possible with an observer-based system ($K_P=1.5$, $K_I=30$, and $K_{DD}=0.0$). Save this plot as red. Finally, repeat for the disturbance-decoupled system ($K_P=1.5$, $K_I=0$, and $K_{DD}=1.0$).
   B. Measure the disturbance response (gain) for all three cases at 3 Hz. (Use the single cursor with the cursor set at 3 Hz; use radio buttons in the cursor display box to set which waveform is measured in the DSA cursor window.)
   C. How much does the disturbance response improve due to increased loop gains allowed by the observer, according to the first two measurements in part A?
   D. How much does the disturbance response improve due to disturbance decoupling, according to second two measurements in part A?
   E. Make similar measurements to part B using the time domain. Set the waveform generator to sine wave (double-click on *WaveGen* block after compiling) and set the frequency to 3 Hz. Repeat for all three gain sets used in part A. Use the main scope block for the most accurate measurements.
   F. Repeat part C using measurements from part E.
   G. Repeat part D using measurements from part E.
   H. Compare measurements made in C and D to those made in F and G.

4. Evaluate the improvement of disturbance response when the power converter bandwidth is increased.
   A. Modify Experiment 6E to view command response. Open Experiment 6E; to avoid permanent changes to the file *Experiment_6E.mqd*, save as *Temp.mqd*. Disconnect the disturbance source from the disturbance input summing junction; disconnect the *Live Constant R* from the command input. Connect the disturbance generator to the command input as shown in Figure 6-22. Tune the nondecoupled system ($K_{DD}=0$) for three power converter bandwidths (FP_C): 50, 100, and 200 Hz. Tuning criteria: no overshoot to step for K_P and 25% overshoot with K_I.

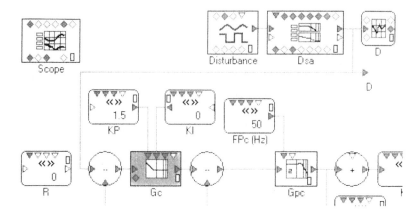

Figure 6-22. Temporarily modified section of Experiment 6E for Exercise 4A.

B. Reopen Experiment 6E. What is the peak excursion of the disturbance response for the three power–converter frequencies without disturbance decoupling, using the tuning gains from part A.

C. Repeat part B with disturbance decoupling ($K_{DD}=1$, $K_I=0$).

D. For the 200-Hz power converter, does raising K_P from the values in part A significantly affect the disturbance response in the disturbance-decoupled system? Explain.

Chapter 7

Noise in the
Luenberger Observer

I n this chapter . . .

- Common sources of noise in control systems
- Effects of noise on observers and observer-based systems
- Effects of noise on disturbance-decoupled systems
- Reducing noise in observer-based systems
- The modified Luenberger observer

This chapter will discuss noise in control systems, especially for systems that use the observed state as a feedback signal. The effect of noise on systems using observer-based disturbance decoupling will also be discussed. In addition, three common methods used to reduce noise sensitivity will be presented: reducing observer bandwidth, filtering the observed disturbance, and modifying the observer-compensator structure. As in earlier chapters, key points will be demonstrated with simulation experiments.

7.1 Noise in Control Systems

Noise sensitivity is an important consideration for most control-systems. Noise from sensors and other sources can distort the control-system output, introducing unwanted perturbations on the control variable and generating unacceptable levels of acoustic noise. An understanding of noise is desirable for designers who use observers because observer-based control systems are often more sensitive to sensor noise than are traditional control systems. This section will provide background on noise, including different types and sources of noise and how noise enters the control system. This background will aid the analytical development that follows in the later sections.

7.1.1 White vs Colored Noise

Random noise is composed of harmonics from across the frequency spectrum. Random noise is often referred to as *white* because, like the color white, it contains all frequency harmonics in approximately equal magnitude. Noise can be filtered, a condition where one or more bands of frequency have been attenuated; this is referred to as *colored* noise. For example, if white noise feeds a low-pass filter, frequency components at and above the filter bandwidth will be attenuated; the output of such a filter will be colored noise.

In this chapter, all analytical and experimental development assumes that the noise is white. The experiments use the unfiltered output of a pseudo-random number generator to produce near-white noise. Bear in mind that noise above the Nyquist frequency (half the sample frequency) aliases to lower frequencies. For those readers studying the effects of colored noise, the results of the treatment of white noise are easily modified to show the effects of colored noise. For analytical treatment, the transfer function of the coloring filter can be cascaded with that of the white noise. For experimental treatment, a coloring filter can be cascaded with the pseudo-random number generator.

7.1.2 Quantization and Noise

Quantization is a common source of noise in digital control systems. Quantization is the undesirable process of limiting resolution of a continuous signal. For example, a 12-bit analog-to-digital converter (ADC) allows only 2^{12} (4096) discrete values to represent a voltage. Even though the voltage input to the ADC almost always falls between these values, it will be assigned one of these discrete values. For the ideal case, the value will be the closest discrete value to the actual value. Assuming that the input (nonquantized) signal can take on any value, quantization is sometimes represented as a random noise added to the actual signal. The magnitude of the random noise is half the resolution of the quantization process. For example, if the ADC were quantized to 0.005 V, the output of the ADC could be modeled as the actual magnitude of the input signal summed with a random noise signal that had a min/max of ±0.0025 V.

Quantization comes from two primary sources. First, in digital control systems, sensor output must be represented digitally. Since sensors usually monitor analog processes, this implies that the sensor output must be quantized. This may occur through standard analog-to-digital converters or through any of the myriad of digital converters for specialized sensors. The second source for quantization is through digital calculations. Many digital calculations generate quantized output

As was the case in Section 7.2, this section will start with an analytical discussion of noise sensitivity based on transfer functions. First, the noise sensitivity of the observed disturbance will be presented; a discussion of noise sensitivity of the actual state will follow that. Finally, the key concepts from this section will be demonstrated in *Visual ModelQ* experiments.

7.3.1.1 Transfer Function Analysis of D_O/N

The noise sensitivity of the observed disturbance can be derived from Figure 7-2 using Mason's signal flow graphs in a manner similar to how Equation 7.1 was derived. The result is

$$D_O(s) = Z(s) \times \frac{G_{CO}(s)}{1 + G_{PEst}(s) \times G_{CO}(s) \times G_{SEst}(s)} - P_C(s) \frac{G_{PEst}(s) \times G_{CO}(s) \times G_{SEst}(s)}{1 + G_{PEst}(s) \times G_{CO}(s) \times G_{SEst}(s)}.$$

$$(7.16)$$

The $Z(s)$ term is multiplied by $G_{PEst}(s) \times G_{PEst}^{-1}(s)$ and by $G_{SEst}(s) \times G_{SEst}^{-1}(s)$ in order to produce a factor of $G_{OLPF}(s)$ (see Equation 7.2). Then, Equation 7.16 can be in a form similar to Equation 7.4:

$$D_O(s) = (Z(s) \times G_{SEst}^{-1}(s) \times G_{PEst}^{-1}(s) - P_C(s)) \times G_{OLPF}(s).   (7.17)$$

The noise sensitivity of the observer can be written in a form similar to Equation 7.5:

$$\frac{D_O(s)}{N(s)} = G_{SEst}^{-1}(s) \times G_{PEst}^{-1}(s) \times G_{OLPF}(s).   (7.18)$$

Comparing the noise sensitivity of the observed state (Equation 7.5) and of the observed disturbance, the difference is that the observed disturbance adds the term $G_{PEst}^{-1}(s)$. If the plant is an integrator, as it commonly is, the inverse is a differentiator, a function well known to be noise sensitive. The more that the simulated plant attenuates noise, the greater the noise sensitivity of the observed disturbance.

7.3.1.2 Transfer-Function Analysis of C/N with Disturbance Decoupling

The sensitivity of the actual state to noise is the final measure on the control system. The next step will be to investigate this by building the transfer function of $C(s)/N(s)$. To simplify transfer-function analysis, the disturbance-decoupled system is drawn in Figure 7-12 by substituting Equation 7.17 in place of the common implementation of the Luenberger observer. Just to be clear, the two implementations differ, but the analysis is valid because transfer functions are the same.

As was the case in Section 7.2, this section will start with an analytical discussion of noise sensitivity based on transfer functions. First, the noise sensitivity of the observed disturbance will be presented; a discussion of noise sensitivity of the actual state will follow that. Finally, the key concepts from this section will be demonstrated in *Visual ModelQ* experiments.

7.3.1.1 Transfer Function Analysis of D_O/N

The noise sensitivity of the observed disturbance can be derived from Figure 7-2 using Mason's signal flow graphs in a manner similar to how Equation 7.1 was derived. The result is

$$D_O(s) = Z(s) \times \frac{G_{CO}(s)}{1 + G_{PEst}(s) \times G_{CO}(s) \times G_{SEst}(s)} - P_C(s)\frac{G_{PEst}(s) \times G_{CO}(s) \times G_{SEst}(s)}{1 + G_{PEst}(s) \times G_{CO}(s) \times G_{SEst}(s)}.$$

$$(7.16)$$

The $Z(s)$ term is multiplied by $G_{PEst}(s) \times G_{PEst}^{-1}(s)$ and by $G_{SEst}(s) \times G_{SEst}^{-1}(s)$ in order to produce a factor of $G_{OLPF}(s)$ (see Equation 7.2). Then, Equation 7.16 can be in a form similar to Equation 7.4:

$$D_O(s) = (Z(s) \times G_{SEst}^{-1}(s) \times G_{PEst}^{-1}(s) - P_C(s)) \times G_{OLPF}(s). \qquad (7.17)$$

The noise sensitivity of the observer can be written in a form similar to Equation 7.5:

$$\frac{D_O(s)}{N(s)} = G_{SEst}^{-1}(s) \times G_{PEst}^{-1}(s) \times G_{OLPF}(s). \qquad (7.18)$$

Comparing the noise sensitivity of the observed state (Equation 7.5) and of the observed disturbance, the difference is that the observed disturbance adds the term $G_{PEst}^{-1}(s)$. If the plant is an integrator, as it commonly is, the inverse is a differentiator, a function well known to be noise sensitive. The more that the simulated plant attenuates noise, the greater the noise sensitivity of the observed disturbance.

7.3.1.2 Transfer-Function Analysis of C/N with Disturbance Decoupling

The sensitivity of the actual state to noise is the final measure on the control system. The next step will be to investigate this by building the transfer function of $C(s)/N(s)$. To simplify transfer-function analysis, the disturbance-decoupled system is drawn in Figure 7-12 by substituting Equation 7.17 in place of the common implementation of the Luenberger observer. Just to be clear, the two implementations differ, but the analysis is valid because transfer functions are the same.

As was the case in Section 7.2, this section will start with an analytical discussion of noise sensitivity based on transfer functions. First, the noise sensitivity of the observed disturbance will be presented; a discussion of noise sensitivity of the actual state will follow that. Finally, the key concepts from this section will be demonstrated in *Visual ModelQ* experiments.

7.3.1.1 Transfer Function Analysis of D_O/N

The noise sensitivity of the observed disturbance can be derived from Figure 7-2 using Mason's signal flow graphs in a manner similar to how Equation 7.1 was derived. The result is

$$D_O(s) = Z(s) \times \frac{G_{CO}(s)}{1 + G_{PEst}(s) \times G_{CO}(s) \times G_{SEst}(s)} - P_C(s) \frac{G_{PEst}(s) \times G_{CO}(s) \times G_{SEst}(s)}{1 + G_{PEst}(s) \times G_{CO}(s) \times G_{SEst}(s)}.$$
$$(7.16)$$

The $Z(s)$ term is multiplied by $G_{PEst}(s) \times G_{PEst}^{-1}(s)$ and by $G_{SEst}(s) \times G_{SEst}^{-1}(s)$ in order to produce a factor of $G_{OLPF}(s)$ (see Equation 7.2). Then, Equation 7.16 can be in a form similar to Equation 7.4:

$$D_O(s) = (Z(s) \times G_{SEst}^{-1}(s) \times G_{PEst}^{-1}(s) - P_C(s)) \times G_{OLPF}(s). \qquad (7.17)$$

The noise sensitivity of the observer can be written in a form similar to Equation 7.5:

$$\frac{D_O(s)}{N(s)} = G_{SEst}^{-1}(s) \times G_{PEst}^{-1}(s) \times G_{OLPF}(s). \qquad (7.18)$$

Comparing the noise sensitivity of the observed state (Equation 7.5) and of the observed disturbance, the difference is that the observed disturbance adds the term $G_{PEst}^{-1}(s)$. If the plant is an integrator, as it commonly is, the inverse is a differentiator, a function well known to be noise sensitive. The more that the simulated plant attenuates noise, the greater the noise sensitivity of the observed disturbance.

7.3.1.2 Transfer-Function Analysis of C/N with Disturbance Decoupling

The sensitivity of the actual state to noise is the final measure on the control system. The next step will be to investigate this by building the transfer function of $C(s)/N(s)$. To simplify transfer-function analysis, the disturbance-decoupled system is drawn in Figure 7-12 by substituting Equation 7.17 in place of the common implementation of the Luenberger observer. Just to be clear, the two implementations differ, but the analysis is valid because transfer functions are the same.

7. Dividing both sides of step 6 by $G_{SEst}(s)$ reveals the problem: $C_O(s) \approx C(s) + N(s) \times G_{SEst}^{-1}(s)$.

This demonstrates that, for frequencies below the observer bandwidth, noise is scaled approximately by $G_{SEst}^{-1}(s)$ and then added to the observed state. Since $G_{SEst}(s)$ typically has the form of a low-pass filter, attenuating frequency components above the sensor bandwidth, $G_{SEst}^{-1}(s)$, will amplify components above the sensor bandwidth.

Another way to see noise amplification in observers is to consider the operation of the observer loop in the presence of noise. Below the observer bandwidth, the observer compensator drives the plant hard enough to remove almost all observer error. Again, $G_{SEst}(s)$ acts like a low-pass filter in practical systems. This means that the noise content above the bandwidth of $G_{SEst}(s)$ must be amplified in order to produce an observed-sensor output ($Y_O(s)$) that will drive the observer error to zero. For example, suppose at 100 Hz, $G_{SEst}(s)$ provides −20 dB (1/10th) gain. Suppose also that $Z(s)$ has a 100-Hz noise component with a magnitude of 0.1 V. Then, the observer compensator would have to drive the estimated plant hard enough that a 100-Hz signal with 1.0-V magnitude appeared at $C_O(s)$. This is required for 0.1 V to exit from $G_{SEst}(s)$ and cancel the 100-Hz noise component in $Z(s)$. So, at 100 Hz, the observed feedback would contain the sensor noise amplified by a factor of ten. Again, the observer compensator effectively amplifies the noise by the $1/G_{SEst}(s)$ or, equivalently, $G_{SEst}^{-1}(s)$.

The noise susceptibility of an observer can be examined more carefully by analyzing the transfer function first discussed in Section 4.3. Recall from Equation 4.4 that the observer output is a combination of the high-pass filtered power converter signal and the low-pass filtered sensor signal. Rewritten with the noisy sensor output, $Z(s)$, Equation 4.4 becomes:

$$C_O(s) = Z(s) \times G_{SEst}^{-1}(s) \frac{G_{PEst}(s) \times G_{CO}(s) \times G_{SEst}(s)}{1 + G_{PEst}(s) \times G_{CO}(s) \times G_{SEst}(s)}$$

$$+ P_C(s) \times G_{PEst}(s) \frac{1}{1 + G_{PEst}(s) \times G_{CO}(s) \times G_{SEst}(s)}. \qquad (7.1)$$

The fraction that appears in the first term of the right side of Equation 7.1 is the closed-loop response of the observer. It behaves like a low-pass filter with a bandwidth equal to that of the observer loop. The fractional portion of the second term on the right side is the high-pass filter first discussed in Section 4.3. Recognizing the low-pass and high-pass filtering terms of Equation 7.1 as

$$G_{OLPF}(s) = \frac{G_{PEst}(s) \times G_{CO}(s) \times G_{SEst}(s)}{1 + G_{PEst}(s) \times G_{CO}(s) \times G_{SEst}(s)} \qquad (7.2)$$

$$G_{OHPF}(s) = \frac{1}{1 + G_{PEst}(s) \times G_{CO}(s) \times G_{SEst}(s)} \qquad (7.3)$$

Eq. (7.1) can be rewritten as:

$$C_O(s) = Z(s) \times G_{SEst}^{-1}(s) \times G_{OLPF}(s) + P_{C'}(s) \times G_{PEst}(s) \times G_{HLPF}(s). \quad (7.4)$$

If the noise content of the power converter, $P_C(s)$, and the nonnoise components of the sensor output are ignored ($Z(s) = N(s)$), noise sensitivity is:

$$\frac{C_O(s)}{N(s)} = G_{SEst}^{-1}(s) \times G_{OLPF}(s). \quad (7.5)$$

Below the observer bandwidth, the term $G_{OPLF}(s)$ is approximately 1. Below the sensor bandwidth, $G_{SEst}(s)$ is approximately 1; above that frequency, $G_{SEst}(s)$ declines. Thus, between the sensor bandwidth and the observer bandwidth, the sensor noise will be amplified.

7.2.1.1 Transfer Function Analysis of C(s)/N(s)

This section will analyze the noise sensitivity of the actual state, $C(s)$. In the end, the sensitivity of the actual state is of more concern than that of the observed state; the observed state is an internal signal while the actual state describes the response of the machine or process.

This analysis begins by considering Figure 7-3, a block diagram of the control system with observed feedback. The observer is represented in the filter form, which was introduced in Section 4.3, using the definitions of Equations 7.2 and 7.3. Evaluation of the noise sensitivity can be carried out by assuming zero command ($R(s) = 0$) and then evaluating the transfer function from the noise, $N(s)$, to the actual state, $C(s)$.

Mason's signal flow graphs can be used to develop the transfer function from $N(s)$ to $C(s)$. In Figure 7-3, there is a single forward path from $N(s)$ to $C(s)$, and there are two loops, which are both in contact with the forward path:

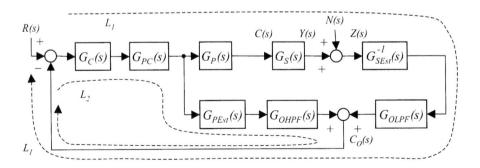

Figure 7-3. Block diagram of system with filter-form representation of observer.

$$P_1 = G_{SEst}^{-1}(s) \times G_{OLPF}(s) \times G_C(s) \times G_{PC}(s) \times G_P(s)$$

$$L_1 = -G_C(s) \times G_{PC}(s) \times G_P(s) \times G_S(s) \times G_{SEst}^{-1}(s) \times G_{OLPF}(s)$$

$$L_2 = -G_C(s) \times G_{PC}(s) \times G_{PEst}(s) \times G_{OHPF}(s).$$

This yields the transfer function of Equation 7.6 or, equivalently, Equation 7.7.

$$T(s) = \frac{P_1}{1 - L_1 - L_2} \qquad (7.6)$$

$$\frac{C(s)}{N(s)} = \frac{G_{SEst}^{-1}(s) \times G_{OLPF}(s) \times G_C(s) \times G_{PC}(s) \times G_P(s)}{1 + G_C(s) \times G_{PC}(s) \times (G_P(s) \times G_S(s) \times G_{SEst}^{-1}(s) \times G_{OLPF}(s) + G_{PEst}(s) \times G_{OHPF}(s))} \qquad (7.7)$$

Equation 7.7 can be simplified by assuming the observer plant and sensor models are nearly ideal and by recognizing that the two filter terms from Equations 7.2 and 7.3 sum to 1:

$$G_{SEst}(s) \approx G_S(s)$$

$$G_{PEst}(s) \approx G_P(s)$$

$$G_{OHPF}(s) + G_{OLPF}(s) = 1.$$

These assumptions simplify Equation 7.7 to:

$$\frac{C(s)}{N(s)} \approx G_{SEst}^{-1}(s) \times G_{OLPF}(s) \times \frac{G_C(s) \times G_{PC}(s) \times G_P(s)}{1 + G_C(s) \times G_{PC}(s) \times G_P(s)}. \qquad (7.8)$$

The result of Equation 7.8 is consistent with Equation 7.5, which demonstrated that the noise sensitivity of the observed state was $G_{SEst}^{-1}(s) \times G_{OLPF}(s)$. That term also appears in the right side of Equation 7.8. Further, notice that the remaining term on the right side of Equation 7.8 is in the form $G_{OL}(s)/(1 + G_{OL}(s))$, where $G_{OL}(s)$ is the open-loop transfer function excluding the effects of the observer and sensor. Finally, recognizing that the system closed-loop transfer function is $G_{OL}(s)/(1 + G_{OL}(s))$, it becomes apparent that the noise sensitivity of the actual state is simply the noise sensitivity of the observed state cascaded with the control-law closed-loop transfer function.

$$\frac{C(s)}{N(s)} \approx G_{SEst}^{-1}(s) \times G_{OLPF}(s) \times G_{CL}(s) \qquad (7.9)$$

7.2.1.2 Comparison to Traditional (Nonobserver) Systems

The noise sensitivity of the traditional and observer-based control systems can be compared by analyzing the differences of their respective transfer functions. The traditional control system with a noisy sensor is shown in Figure 7-4.

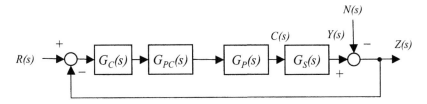

Figure 7-4. Block diagram of traditional (nonobserver) system.

The noise sensitivity of the actual state, $C(s)$, is written out from Mason's signal flow graphs:

$$\frac{C(s)}{N(s)} = \frac{G_C(s) \times G_{PC}(s) \times G_P(s)}{1 + G_C(s) \times G_{PC}(s) \times G_P(s) \times G_S(s)}. \qquad (7.10)$$

Rearranging Equation 7.10 to isolate the closed-loop transfer function[1] yields:

$$\frac{C(s)}{N(s)} = \frac{G_C(s) \times G_{PC}(s) \times G_P(s)}{1 + G_C(s) \times G_{PC}(s) \times G_P(s) \times G_S(s)} \qquad (7.11)$$

$$\frac{C(s)}{N(s)} = G_S^{-1}(s) \times G_{CL}(s).$$

At frequencies well below the control-law bandwidth, the open-loop gain will dominate the "1" in the denominator and the noise susceptibility will be:

$$\frac{C(s)}{N(s)} \approx G_S^{-1}(s). \qquad (7.12)$$

A similar result occurs when the observer-based transfer function of Equation 7.8 is evaluated below the observer bandwidth (where $G_{OLPF}(s) \approx 1$) and the control-law bandwidth (where the closed-loop control-law response ≈ 1). Equation 7.8 reduces to:

$$\frac{C(s)}{N(s)} \approx G_{SEst}^{-1}(s). \qquad (7.13)$$

---

[1]The reader should notice that the forms of the open-loop transfer functions differed between the traditional and observer-based feedback systems. The open loop of the traditional system includes the sensor transfer function while the observer-based system does not. The reason for the difference is that the observer removes the effects of the sensor transfer function when properly configured and thus its elimination accurately reflects the operation of the control loop.

So, at low frequencies, the noise susceptibility of the two systems is about the same. However, there is a significant difference when the transfer functions are evaluated at higher frequencies. In the traditional system, at frequencies higher than the control-law bandwidth, the "1" in the denominator of Equation 7.11 dominates and the noise sensitivity reduces to:

$$\frac{C(s)}{N(s)} = G_C(s) \times G_{PC}(s) \times G_P(s). \qquad (7.14)$$

However, the observer-based control system produces a different result. Above the control-law bandwidth, where "1" dominates the denominator of Equation 7.8, that equation reduces to:

$$\frac{C(s)}{N(s)} = G_{SEst}^{-1}(s) \times G_{OLPF}(s) \times G_C(s) \times G_{PC}(s) \times G_P(s). \qquad (7.15)$$

Comparing the noise susceptibility indicated by Equations 7.14 and 7.15, the primary difference is the appearance of the term $G_{SEst}^{-1}(s)$ in the observer-based system. Since $G_{SEst}(s)$ is normally a term that attenuates at high frequency, $G_{SEst}^{-1}(s)$ will normally amplify at high frequency. Above the sensor bandwidth and below the observer bandwidth, the observer-based system will be noisier by an amount approximately equal to the attenuation provided by the sensor.

7.2.2 Experiment 7A: The Effects of Sensor Noise on the Observed State

This section and the next will use simulations to confirm the previous analytical development. The simulations will start with *Visual ModelQ* Experiment 7A, shown in Figure 7-5. This model is similar to the models of previous chapters. The key points are:

- The control system is a PI loop with the feedback taken from the sensor. Similar to Experiments 5C and 6A, the sensor is modeled as a single-pole low-pass filter with a bandwidth of 20 Hz, and the power converter is a double-pole low-pass filter with a bandwidth of 50 Hz. The PI gains are also the same as in those experiments.
- There is an observer that is tuned to a bandwidth of about 155 Hz (see Section 5.3.1). The model parameters of the observer are accurate: $K_{Est} = 50$ and $F_{GsEst} = 20$ Hz.
- The model includes a noise source named "Sensor Noise" near the top center. This is a pseudorandom number generator with an amplitude of ±1. The sum of the noiseless sensor output and the noise source form Z, the sensor output.
- There is a DSA configured to inject the noise input of the control loop. This instrument is used to measure the noise response of various signals in the

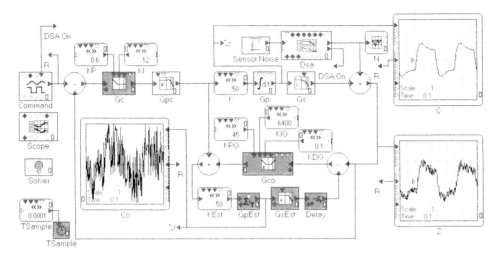

Figure 7-5. Experiment 7A: Evaluating the effects of noise on the observed state.

system. Note that the DSA disables the command waveform generator (at left) through the *Visual ModelQ* Extender "DSA On." This is done because the DSA must zero other control-system inputs during its excitation to allow accurate measurements. As usual, when the DSA is not exciting the system (the normal case), it passes the output of the random generator Sensor Noise to the summing junction that adds Y and N to form Z.

- There are three *Live Scopes*: the actual state (C), the measured state after noise is injected (Z), and the observed state (C_O).

The results of Experiment 7A can be viewed in the time domain and in the frequency domain. The time domain aids intuitive understanding of the observer behavior in the presence of noise; the frequency domain provides quantitative comparisons.

The result in the time domain can be seen in the *Live Scopes* of Figure 7-5. The actual state, C, at top right, is almost unaffected by noise in this system. That is because the control-law gains are relatively low and the integrating plant provides enough filtering to remove visible effects of the noise source. The effects of the noise on the sensor output, Z, at bottom right, are apparent. Without the noise, Z would be nearly identical to C, except for the phase lag of the sensor, $G_S(s)$.

The observed state, C_O, at left, is much more affected than is the sensor output, Z. In the absence of noise, C_O is almost identical to C (reference Figure 5-5 from Experiment 5C). Of course, these signals are not the same in the presence of noise. The noise, fed in through Z, has been amplified by the observer. The result is that the observed state is corrupted by the noise to such an extent that the signal actual state is difficult to see through the high-frequency noise. This is the expected result from Equation 7.5.

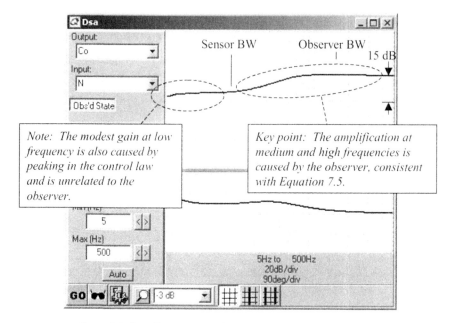

Figure 7-6. From Experiment 7A: Bode plot of observed state vs noise.

A Bode plot of observed state vs noise, as shown in Figure 7-6, provides a quantitative evaluation of the effects of noise. In the middle frequencies (approximately between the sensor bandwidth and the observer bandwidth) the gain is steadily rising. This can be seen in Equation 7.5 because the amplification of $G_{SEsl}^{-1}(s)$ increases above the sensor bandwidth. The level of amplification stabilizes around the observer bandwidth. This can also be seen in Equation 7.5 because, above the observer bandwidth, the observer (represented at $G_{OLPF}(s)$) provides attenuation.

The reader may notice an unexpected amplification of a few decibels below the sensor bandwidth. This is not predicted by Equation 7.5. In fact, this results because of peaking in the control loop; the cause of this effect is that the control system has some peaking in this range so that noise signals, as well as commands, are modestly amplified in this frequency range.

7.2.3 Experiment 7B: The Effects of Sensor Noise on the Actual State

Experiment 7B extends 7A to allow the evaluation of the effects of noise on the actual state (see Figure 7.7). There are just a few differences:

- The feedback path is fed with a switch (at bottom left) that can be configured to connect either sensor output or the observed state. The connection is the

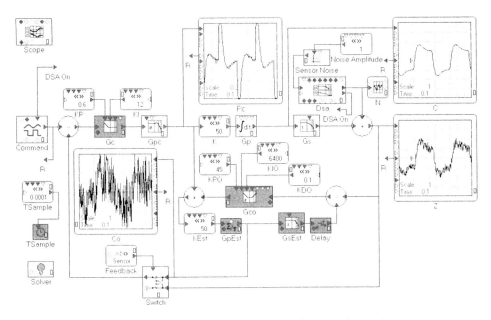

Figure 7-7. Experiment 7B: Evaluating the effects of noise on the actual state.

controller with the *Live Constant* "Feedback," which can take on the values of "Sensor" or "Observer."

- The amplitude of the noise is controlled with a *Live Constant* named "Noise Amplitude" at top right.
- The power converter output, P_C, is shown in a *Live Scope* at top center. The power converter output is an indicator of noise susceptibility.

The effects of noise in the time domain are most easily seen by viewing the power converter output. This is because an integrating plant smoothes the noise, making evaluation based on the actual state difficult. Also, many noise issues are driven by the power converter output rather than by the actual state. For example, some noise in a hydraulic positioning system will be caused by the flow of hydraulic fluid rather than by the position of the cylinder. Acoustic noise from a voltage supply will often be caused by current that vibrates inductor winding, not by the output voltage.

The comparison of the power converter output of Experiment 7B is given in Figure 7-8. In Figure 7-8a, where the sensor output is used for feedback, the effects of noise are modest. In Figure 7-8b, where the observed state is used for feedback, the effects are considerably greater. Note that the noise amplitude and control gains (K_P, K_I) are identical in these plots; the increase in noise sensitivity is due solely to the use of the observed state.

The noise sensitivity of the actual state also can be viewed by comparing the actual state with the loop configured for (1) observed-state feedback and (2) sensor output.

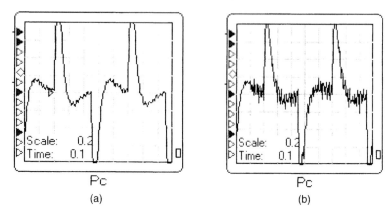

Figure 7-8. From Experiment 7B: Effects of noise on the power converter output (a) with sensor output as feedback and (b) with observed state for feedback.

A comparison is shown in Figure 7-9. Again, the result is that the sensitivity of the system with observer feedback (Figure 7-9a) is greater than that of the system with sensor feedback (Figure 7-9b). The integrating plant smoothes noise, making it difficult to discern in these plots. In fact, the amplitude of the noise source had to be increased from 1 to 5 to show the effects even this much. The plots of Figure 7-9 are consistent with Equations 7.14 and 7.15; the effect of the term $G_{SESi}^{-1}(s) \times G_{OLPF}(s)$ in Equation 7.15 will normally amplify high-frequency noise.

A quantitative measurement of noise sensitivity of the actual state can be seen in a Bode plot generated by Experiment 7B and shown in Figure 7-10. The noise sensitivity of the sensor-based system (Equation 7.14) and the observed-state-based

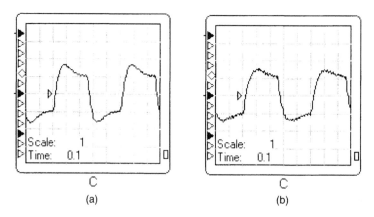

Figure 7-9. From Experiment 7B: Effects of noise on the actual state (a) with sensor output as feedback and (b) with observed state for feedback. Note that the noise amplitude has been increased to 5 for both cases.

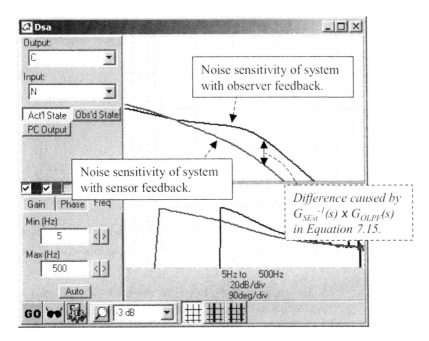

Figure 7-10. From Experiment 7B: Bode plot of noise sensitivity of actual state with observer feedback and sensor feedback.

system (Equation 7.15) is plotted in this figure. The difference between the two configurations is due to the term $G_{SEst}^{-1}(s) \times G_{OLPF}(s)$, which is the difference between Equations 7.14 and 7.15. That term is the noise sensitivity of the observed state; it is equal to Equation 7.5 and is plotted approximately in Figure 7-6. As expected, the observed-state-based system is equivalent to the sensor-based system at frequencies below the sensor bandwidth, but is more sensitive above that. The maximum difference of about 15 dB is present at and above the observer bandwidth, consistent with Figure 7-6.

7.2.4 The Effects of Control-Law Gains on Noise Sensitivity

Equation 7.15 predicts that noise sensitivity of the observer-based system will increase with increased control-law gains (G_C). The comparisons of the previous sections assumed the control-law gains were the same with and without the observer. However, observers are often employed to allow increased control-law gains. Since the control-law gains will often be higher in observer-based systems, noise sensitivity will increase further.

The effect of raising control-law gains is shown in Figure 7-11. In Figure 7-11a, the sensitivity with the original control-law gains ($K_P=0.6$, $K_I=12$) is shown; this is

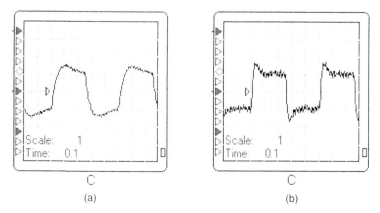

(a)   (b)

Figure 7-11. From Experiment 7B, the effects of control law gains on sensitivity of actual state in observer-based system: (a) with low gains ($K_P=0.6$, $K_I=12$), and (b) with high gains ($K_P=1.5$, $K_I=30$).

identical to Figure 7-9b. In Figure 7-11b, taking advantage of the reduction in phase lag provided by the observer, the gains are raised to ($K_P=1.5$, $K_I=30$), which are the higher gains used in Chapter 5. The increase in sensitivity is apparent. This is consistent with Equation 7.15, which shows noise sensitivity increasing with control-law gains, as well as with other factors including the observer and power converter bandwidths.

7.3 Noise Sensitivity when Using Disturbance Decoupling

Disturbance decoupling can greatly increase the noise sensitivity of a control system for at least two reasons. First, as demonstrated in Section 6.3, disturbance-decoupled systems can respond at higher frequencies than the control law. The structure of the disturbance-decoupled system routes the observed disturbance directly to the power converter; the primary limits on response are the observer bandwidth, the power converter bandwidth, and the plant. This structure gives disturbance decoupling its superior disturbance-rejection properties. Unfortunately, it brings with it higher, often much higher, noise sensitivity.

The second reason for higher noise sensitivity is that, where observer-feedback systems rely on the observed state, disturbance decoupling relies on the observed disturbance. The observed disturbance is more sensitive to noise than the observed state. In this context, the main difference between the two signals is that the observed state passes through the model plant (reference Figure 7-2). In practical systems, the model plant usually will be a combination of integrators and low-pass filters, both of which will attenuate noise. For disturbance-decoupled systems, the unattenuated output of the observer compensator is routed directly to the power converter, often increasing the noise sensitivity by many times.

As was the case in Section 7.2, this section will start with an analytical discussion of noise sensitivity based on transfer functions. First, the noise sensitivity of the observed disturbance will be presented; a discussion of noise sensitivity of the actual state will follow that. Finally, the key concepts from this section will be demonstrated in *Visual ModelQ* experiments.

7.3.1.1 Transfer Function Analysis of D_O/N

The noise sensitivity of the observed disturbance can be derived from Figure 7-2 using Mason's signal flow graphs in a manner similar to how Equation 7.1 was derived. The result is

$$D_O(s) = Z(s) \times \frac{G_{CO}(s)}{1 + G_{PEst}(s) \times G_{CO}(s) \times G_{SEst}(s)} + P_C(s) \frac{G_{PEst}(s) \times G_{CO}(s) \times G_{SEst}(s)}{1 + G_{PEst}(s) \times G_{CO}(s) \times G_{SEst}(s)}.$$

$$(7.16)$$

The $Z(s)$ term is multiplied by $G_{PEst}(s) \times G_{PEst}^{-1}(s)$ and by $G_{SEst}(s) \times G_{SEst}^{-1}(s)$ in order to produce a factor of $G_{OLPF}(s)$ (see Equation 7.2). Then, Equation 7.16 can be in a form similar to Equation 7.4:

$$D_O(s) = (Z(s) \times G_{SEst}^{-1}(s) \times G_{PEst}^{-1}(s) + P_C(s)) \times G_{OLPF}(s). \qquad (7.17)$$

The noise sensitivity of the observer can be written in a form similar to Equation 7.5:

$$\frac{D_O(s)}{N(s)} = G_{SEst}^{-1}(s) \times G_{PEst}^{-1}(s) \times G_{OLPF}(s). \qquad (7.18)$$

Comparing the noise sensitivity of the observed state (Equation 7.5) and of the observed disturbance, the difference is that the observed disturbance adds the term $G_{PEst}^{-1}(s)$. If the plant is an integrator, as it commonly is, the inverse is a differentiator, a function well known to be noise sensitive. The more that the simulated plant attenuates noise, the greater the noise sensitivity of the observed disturbance.

7.3.1.2 Transfer-Function Analysis of C/N with Disturbance Decoupling

The sensitivity of the actual state to noise is the final measure on the control system. The next step will be to investigate this by building the transfer function of $C(s)/N(s)$. To simplify transfer-function analysis, the disturbance-decoupled system is drawn in Figure 7-12 by substituting Equation 7.17 in place of the common implementation of the Luenberger observer. Just to be clear, the two implementations differ, but the analysis is valid because transfer functions are the same.

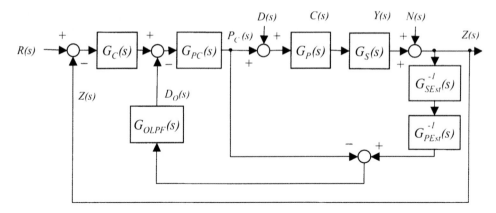

Figure 7-12. Noise in an observer-based disturbance-decoupled system redrawn according to Equation 7.17.

The transfer function from $N(s)$ to $C(s)$ can be written using Mason's signal flow graphs. There are two paths from $N(s)$ to $C(s)$, one through the observer (P_1) and the other through the control law (P_2). There are three loops: one through the control law (L_1) and two passing through the observer (L_2 and L_3).

$$P_1 = G_{SEst}^{-1}(s) \times G_{PEst}^{-1}(s) \times G_{OLPF}(s) \times G_{PC}(s) \times G_P(s) \tag{7.19}$$

$$P_2 = G_C(s) \times G_{PC}(s) \times G_P(s) \tag{7.20}$$

$$L_1 = -G_C(s) \times G_{PC}(s) \times G_P(s) \times G_S(s) \tag{7.21}$$

$$L_2 = -G_{PC}(s) \times G_P(s) \times G_S(s) \times G_{SEst}^{-1}(s) \times G_{PEst}^{-1}(s) \times G_{OLPF}(s) \tag{7.22}$$

$$L_3 = G_{PC}(s) \times G_{OLPF}(s) \tag{7.23}$$

Since all loops touch (via $G_{PC}(s)$), the denominator of the transfer function contains no combinations of loops. Since all forward paths touch all loops (again, via $G_{PC}(s)$), the cofactors, which appear in the numerator, are all 1. Thus, the transfer function is simply:

$$\frac{C(s)}{N(s)} = \frac{P_1 + P_2}{1 - L_1 - L_2 - L_3}. \tag{7.24}$$

Two assumptions simplify the transfer-function analysis. Assuming that the estimated plant and estimated sensor are accurate, ($G_P(s) \approx G_{PEst}(s)$ and $G_S(s) \approx G_{SEst}(s)$), L_2 and L_3 cancel in the denominator. Also, Equation 7.19 simplifies to:

$$P_1 \approx G_{SEst}^{-1}(s) \times G_{OLPF}(s) \times G_{PC}(s). \tag{7.25}$$

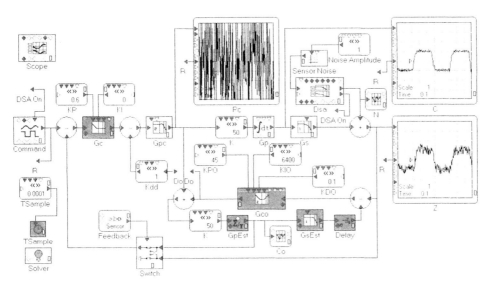

Figure 7-13. Experiment 7C: Noise in an observer-based disturbance-decoupled system.

The transfer function of noise sensitivity is then:

$$\frac{C(s)}{N(s)} \approx \frac{G_{SEst}^{-1}(s) \times G_{OLPF}(s) \times G_{PC}(s) + G_C(s) \times G_{PC}(s) \times G_P(s)}{1 + G_C(s) \times G_{PC}(s) \times G_P(s) \times G_S(s)}. \quad (7.26)$$

Equation 7.26 indicates that the disturbance-decoupled system has much greater noise sensitivity than does the system with observer feedback (Equation 7.8). The denominators of Equations 7.8 and 7.26 are similar, but the first term in the numerator of Equation 7.26, which represents the decoupling path from noise to the actual state, bypasses the control law.[2] The amplifying factor ($G_{SEst}^{-1}(s)$) lacks the attenuating term $G_{PEst}(s)$. In practice, the noise passing through the disturbance-decoupling path, P_1, will normally be much larger than the noise passing through the control-law path, P_2. (In fact, in many cases, P_2 can be ignored.) This accounts for the disturbance-decoupled system's greatly increased noise sensitivity.

7.3.2 Experiment 7C: Noise Susceptibility and Disturbance Decoupling

This section will use Experiment 7C, shown in Figure 7-13, to demonstrate the noise sensitivity of observer-based disturbance decoupling. This experiment is the same as Experiment 7B with two exceptions. First, disturbance decoupling has

---

[2] The second term in the numerator represents the disturbance response of the control law based on measured feedback.

been added using the extenders named D_O to connect the output of the observer compensator to the decoupling gain, K_{DD}. The extender is preferred to a direct connection because it makes the diagram clearer by avoiding crossing wires. Second, the *Live Scope* display of C_O has been removed; a variable block, C_O, has been added so the observed state can be plotted in the DSA and main scope. The greatly increased noise sensitivity of the decoupled system is made readily apparent by the graph of P_C, the power converter output, which is much noisier than it was in any of the nondecoupled systems in earlier models (for example, in Figure 7-8).

The Bode plot of Experiment 7C shown in Figure 7-14 provides a measure of the decoupled system's noise sensitivity. This figure shows three configurations. The traditional (nonobserver) configuration has the lowest sensitivity. The system with observer feedback is more sensitive, as was demonstrated in Figure 7-10. However, the disturbance decoupled system is the most sensitive. For example, at 100 Hz, the decoupled system is 25 dB (about 20 times) more sensitive than the observer-based system and more than 40 dB (100 times) more sensitive than the traditional system. Clearly, any consideration of observer-based disturbance decoupling should include a thorough evaluation of noise sensitivity.

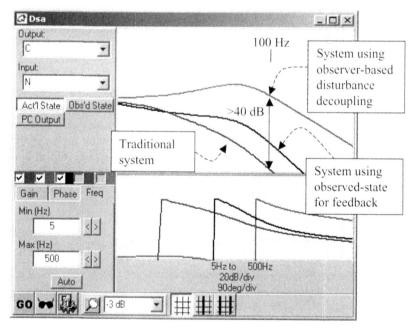

Figure 7-14. From Experiment 7C, noise sensitivity of the actual state for three configurations: traditional, the observed-state feedback, and the observer-based disturbance-decoupled system.

7.4 Reducing Noise Susceptibility in Observer-Based Systems

This section will review several techniques used to reduce noise susceptibility in observer-based systems.

7.4.1 Lowering Observer Bandwidth

One technique to reduce noise sensitivity is to lower the observer bandwidth. While the tuning procedures covered in earlier chapters focused on helping the designer tune the compensator to get the maximum bandwidth available from the observer, in practical systems the observer's target bandwidth may be reduced to attenuate noise.

Reducing the observer bandwidth will result in a direct benefit for noise susceptibility. This is apparent upon inspection for systems using observed-state feedback from Equation 7.9, understanding that reducing observer bandwidth is equivalent to reducing the bandwidth of the equivalent filter $G_{OLPF}(s)$. It is also true for disturbance-decoupled systems according to Equation 7.26, assuming the first term in the numerator of the right-hand side is dominant, as it normally will be.

The effect of lowering observer bandwidth is shown in Figure 7-15. This Bode plot shows the noise susceptibility of the observer-based system (without disturbance

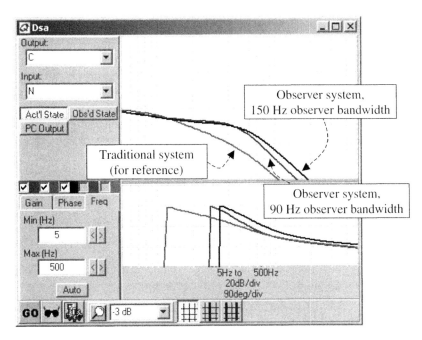

Figure 7-15. From Experiment 7B: Comparing the noise susceptibility of the observer-based system with high- and low-observer bandwidth.

decoupling) of Experiment 7C. There are three plots: (1) the 155-Hz observer loop used through most of this chapter, (2) that same system with the observer loop tuned down to 90 Hz, and (3) the traditional system for reference. The observer-based system compensated for 155 Hz (K_{DO}=0.1, K_{PO}=45, K_{IO}=6400) has the highest noise susceptibility. The observer was tuned down to 90 Hz (K_{DO}=0.05, K_{PO}=20, K_{IO}=1400) using the procedure from Section 5.3.1, with K_{DO} set to 0.05 to reduce noise. Tuning down the observer provided a modest reduction of 5 dB (the ratio of the two observer bandwidths, 90/155) as is predicted by Equation 7.9.

The problem with reducing observer bandwidth is the system disturbance response may worsen. For systems using the observed state for feedback, this is a concern when the observer is near or below the control-loop bandwidth, as discussed in Section 6.2.2. The reason is that the disturbance enters the control law only after passing through the observer. Low-observer bandwidth will delay the perturbations caused by the disturbance entering the observed state, which is the information the control law has concerning the disturbance.

7.4.2 Reducing Noise in Disturbance-Decoupled Systems

Lowering the observer bandwidth will produce a similar benefit for systems with disturbance decoupling, as is predicted by Equation 7.26. However, lowering the observer bandwidth will degrade the disturbance response in the case where the observer bandwidth is lower than the power converter bandwidth, as shown in Equation 6.15. The primary limits on the disturbance response are the observer bandwidth, which delays the disturbance signal, and the power converter, which delays the decoupling signal entering the control system. The slower of the two will be the primary limit. Lowering the primary limit will degrade the disturbance response.

An alternative means for reducing noise sensitivity when using disturbance decoupling is to add a low-pass filter in line with K_{DD}, as in Experiment 7D as shown in Figure 7-16. The low-pass filter bandwidth is set with the *Live Constant DD LPF*. The improvement of noise susceptibility from the low-pass filter is demonstrated in Figure 7-17. When the filter is reduced from 250 to 50 Hz, the noise is attenuated accordingly.

The degradation of disturbance response from the low-pass filter is shown in Figure 7-18. The response to disturbances worsens (grows) with lower bandwidth filtering on the observed disturbance. However, as long as the filter bandwidth is more than about four times the power converter bandwidth, lowering the filter bandwidth will have a negligible effect on disturbance response.

Adding a low-pass filter in line with the observed disturbance provides an observed disturbance signal similar to that which would have been produced by lowering the observer bandwidth. The primary benefit to the additional filter compared to reducing observer bandwidth is that the filter can be used to lower the bandwidth for decoupling, where noise susceptibility is so great, without reducing the observer bandwidth. If observed-state feedback is used for other purposes, the benefits of

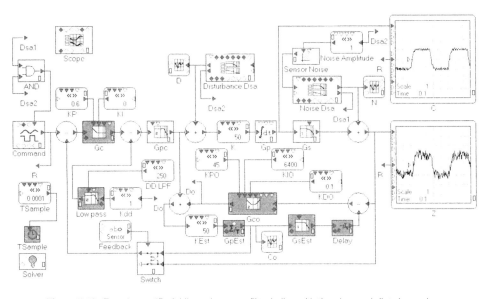

Figure 7-16. Experiment 7D: Adding a low-pass filter in line with the observed disturbance in a disturbance-decoupled system.

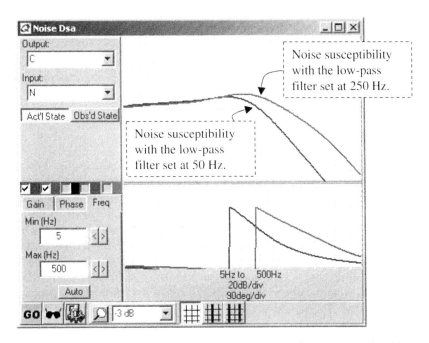

Figure 7-17. Noise susceptibility with the disturbance-decoupling filter set at 50 and 250 Hz.

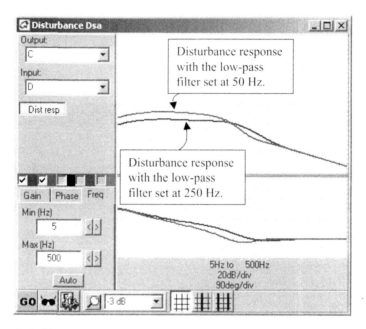

Figure 7-18. Disturbance response with the disturbance-decoupling filter set at 50 and 250 Hz.

high-observer bandwidth for that signal can be maintained. Since the noise sensitivity of the observed-state feedback is so much less than that of the observed disturbance, having the extra degree of design freedom brought by the decoupling filter will be of benefit in some of the applications where both signals from the observer are used simultaneously.

7.4.3 Modifying the Observer Compensator

Noise sensitivity can be reduced by modifying the observer compensator. This method divides the observer compensator into two paths: high frequency and low frequency. The high-frequency path is routed so that it does not contribute noise to the observed disturbance or, optionally, to the observed state. (This is equivalent to the velocity observer of [38, Figure 9].) This method provides similar benefits to reducing observer bandwidth and to filtering the observed disturbance. The modified observer compensator uses fewer computational resources than explicitly filtering the observed disturbance as shown in Figure 7-16. Like Figure 7-16, the restructured compensator has the benefit of allowing the observed disturbance signal to be more heavily filtered than the observed state.

7.4.3.1 Modified Structure

Modifying the observer compensator requires the following steps:

1. Divide $G_{CO}(s)$ into two components, a low-frequency component, $G_{COL}(s)$, and a high-frequency component, $G_{COH}(s)$. Construct $G_{COH}(s)$ so that when multiplied by the simulated plant the product is a simple function such as a constant. The two paths must sum to the original compensator:

$$G_{CO}(s) = G_{COL}(s) + G_{COH}(s).$$

For example, for an integrating plant, the D-path of the observer compensator could form $G_{COH}(s)$ while the I-path and P-path would form $G_{COL}(s)$. Here, when $G_{COH}(s)$ $(K_{DO} \times s)$ is multiplied by the plant (K_{Est}/s), the result is the constant $K_{DO} \times K_{Est}$, which certainly qualifies as a simple function.
2. Form the observed disturbance with the $G_{COL}(s)$ path.
3. Form the observed state with the $G_{COL}(s)$ path. An alternative is to form the observed state with both paths of the observer compensator. Using this alternative will allow the equivalent of filtering the observed disturbance without affecting the observed state.

The modified observer is shown in Figure 7-19. The compensator is divided in two and each path proceeds through an independent simulated plant. As stated above, the

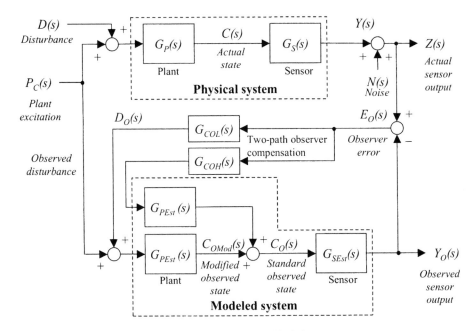

Figure 7-19. Luenberger observer with modified observer compensator.

product of $G_{COH}(s)$ and $G_{PEst}(s)$ should be a simple function such as a constant. In the actual implementation, only the product of the two is evaluated to reduce computations. A common case is where the estimated plant is fully integrating (K_{Est}/s), and $G_{COH}(s)$ is the D-path ($K_{DO} \times s$) of the observer compensator. Here, the path through $G_{COH}(s)$ and $G_{PEst}(s)$ is evaluated in one step as $K_{Est} \times K_{DO}$.

Note that there are two choices for the observed state. The modified observed state, $C_{OMod}(s)$, is derived using only $G_{OL}(s)$. The standard observed state, $C_O(s)$, is constructed using the sum of $G_{OH}(s)$ and $G_{OL}(s)$. It is a straightforward matter to show that $C_O(s)$ in Figure 7-19 is equivalent to $C_O(s)$ in the traditional Luenberger observer used throughout this book.

7.4.3.2 Modifying the Compensator for Nonintegrating Plants

If the plant is something other than a pure integration, this method can still be applied. For example, suppose the plant is a leaky integrator such as a temperature bath that leaks heat to the atmosphere or a motion system with viscous damping. Such a form is equivalent to a scaled low-pass filter as shown in Equation 7.27. Here K_{Leak} is one over the time constant of the leakage:

$$G_{PEst}(s) = \frac{K_{Est}}{s + K_{Leak}}. \tag{7.27}$$

Since the plant is not a pure integration, modifying the observer compensator with $G_{COH}(s) = K_{DO} \times s$ fails to meet the criterion where $G_{COH}(s) \times G_{PEst}(s)$ is to be a simple function such as a constant. However, if a portion of the compensator's proportional gain is placed into $G_{COH}(s)$, then the criterion is met: $G_{COH}(s) = (1 + K_{Leak}/s) \times K_{DO} \times s$. Since $G_{COH}(s)$ now contains a portion of the proportional gain ($K_{Leak} \times K_{DO}$), the proportional gain in $G_{COH}(s)$ must be reduced by that same amount. So,

$$G_{COH}(s) = \left(1 + \frac{K_{Leak}}{s}\right) K_{DO} \times s \tag{7.28}$$

$$G_{COH}(s) \times G_{PEst}(s) = K_{Est} \times K_{DO} \tag{7.29}$$

$$G_{COL}(s) = \frac{K_{IO}}{s} + (K_{PO} - K_{Leak} \times K_{DO}) \tag{7.30}$$

$$G_{CO}(s) = G_{COH}(s) + G_{COL}(s) = \frac{K_{IO}}{s} + K_{PO} + K_{DO} \times s. \tag{7.31}$$

In this case, the two compensator paths still sum to the original $G_{CO}(s)$ as indicated by Equation 7.31. In addition, the evaluation of the high-frequency path observer is still computationally efficient owing to Equation 7.29. The only complexity introduced by the nonintegrating plant is that the proportional gain of the compensator must be reduced as indicated in Equation 7.30.

7.4.4 Transfer-Function Analysis

The transfer-function analysis provides an analytical understanding of modifying the observer compensator. The observed state and observed disturbance can be evaluated separately. Using Mason's signal flow graphs, the observed state can be written as:

$$\frac{C_O(s)}{N(s)} = \frac{G_{COL}(s) \times G_{PEst}(s)}{1 + (G_{COH}(s) \times G_{PEst}(s) + G_{COL}(s) \times G_{PEst}(s)) \times G_{SEst}(s)}. \quad (7.32)$$

Making use of Equation 7.31, Equation 7.32 simplifies to:

$$\frac{C_O(s)}{N(s)} = \frac{G_{COL}(s) \times G_{PEst}(s)}{1 + G_{CO}(s) \times G_{PEst}(s) \times G_{SEst}(s)}. \quad (7.33)$$

Using some algebra, Equation 7.33 can be written as:

$$\frac{C_O(s)}{N(s)} = \frac{G_{COL}(s)}{G_{CO}(s)} \times G_{SEst}^{-1}(s) \times \frac{G_{SEst}(s) \times G_{CO}(s) \times G_{PEst}(s)}{1 + G_{CO}(s) \times G_{PEst}(s) \times G_{SEst}(s)}. \quad (7.34)$$

Recalling the formula for $G_{OLPF}(s)$ as shown in Equation 7.2, Equation 7.34 reduces to

$$\frac{C_O(s)}{N(s)} = \frac{G_{COL}(s)}{G_{CO}(s)} \times G_{SEst}^{-1}(s) \times G_{OLPF}(s). \quad (7.35)$$

Comparing Equation 7.35 to the standard sensitivity as shown in Equation 7.5, the effect of modifying the observer amounts to cascading the transfer function $G_{COL}(s)/G_{CO}(s)$. Since $G_{COL}(s)$ is the low-frequency path of $G_{CO}(s)$, the term $G_{COL}(s)/G_{CO}(s)$ attenuates higher frequencies. In other words, the modified compensator is effectively a low-pass filter. The same effect can be seen when evaluating the observed disturbance. Using a similar technique to that above, it can be shown that

$$\frac{D_O(s)}{N(s)} = \frac{G_{COL}(s)}{G_{CO}(s)} \times G_{SEst}^{-1}(s) \times G_{PEst}^{-1}(s) \times G_{OLPF}(s). \quad (7.36)$$

Comparing Equation 7.36 to the standard observer compensator of Equation 7.18, the same filtering term ($G_{COL}(s)/G_{CO}(s)$) appears with the modified compensator. So, for both outputs of the observer, the difference between the standard and the modified compensator is the introduction of a filtering term.

7.4.5 Experiment 7E: Evaluating the Modified Observer Compensator

Experiment 7E, shown in Figure 7-20, is a system with an observer that can be configured to feed back the standard or modified observed state. The modified observer has a $G_{COII}(s)$ path using K_{DO} and a K_{Est} block and a $G_{COII}(s)$ path using G_{CO} and a second K_{Est} block. The control loop can be configured for one of three feedback signals: (1) the sensor, (2) the standard observed state, and (3) the modified observed state; the configuration is selected with the *Live Constant* "Feedback" which can take on the values of "*Sensor*," "*Observer*," and "*Mod. Obs.*" The selected feedback signal connects to the variable "*F.*"

Results of Experiment 7E are shown in Figure 7-21. The noise susceptibility of the standard and modified observed states is plotted. As expected, the modified observed state is similar to the standard observed state at low frequencies but is attenuated above a frequency (about 70 Hz). This is consistent with the modified observed state being a filtered version of the standard observed state, as shown in Equation 7.35.

Incidentally, the disturbance response does not suffer measurably from the use of the modified observed state, because 70 Hz is still much greater than the control-loop bandwidth (about 25 Hz, from Figure 5-9). As demonstrated in Equation 6.14, in the event of a wide separation of the control-loop bandwidth and the observer bandwidth, the disturbance response of nondisturbance-decoupled systems will be limited by the lesser of the two.

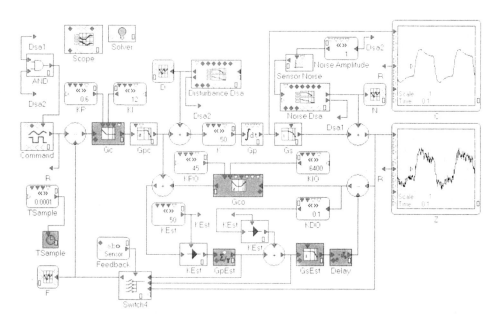

Figure 7-20. Experiment 7E: Modified observer compensator.

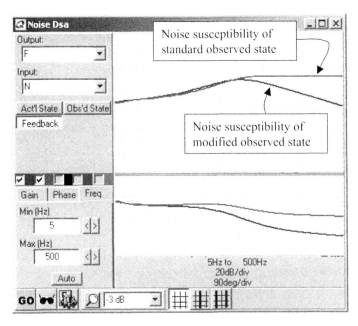

Figure 7-21. From Experiment 7E: Noise susceptibility of feedback signal, with system configured for standard and modified observed state feedback.

7.4.6 Using the Modified Observer Solely for D_O

One feature of the modified observer is that it can be applied solely to the observed disturbance. Referring to Figure 7-19, if the standard observed state, $C_O(s)$, is used for feedback, only the observed disturbance will be affected by the modified compensator. The benefit of this structure is that the observed disturbance is more heavily filtered than the observed state. As demonstrated in the analysis and simulations throughout this chapter, the observed disturbance is much more sensitive to noise than is the observed state. Thus by taking the observed disturbance from the right side of the summing junction, the implicit filtering offered by the modified compensator is applied where it is most needed: to the observed disturbance. This will, in some cases, eliminate the need to provide an explicit filter on the observed disturbance as shown in Figure 7-16. For a system simultaneously using the observed disturbance and the observed state, reducing the observer bandwidth still can be used to reduce the sensitivity of both signals simultaneously. In this case, the observer-compensator gains could be reduced until the sensitivity of the actual state is acceptable; then the modified compensator could be applied to reduce the sensitivity of the observed disturbance.

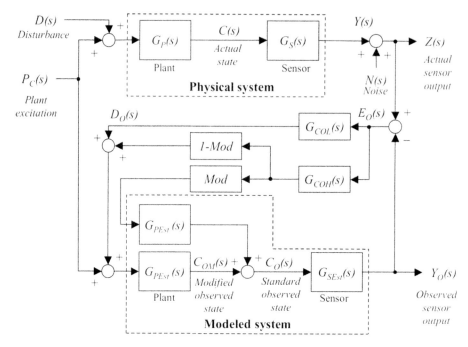

Figure 7-22. Gradually modified observer.

7.4.7 Modifying the Observer Compensator Gradually

This choice of using the modified observer at first appears to lack the option of partial implementation. The designer can use the standard or the modified structure, but nothing in between. However, the modified observer structure of Figure 7-19 can be implemented gradually by routing only a portion of $G_{COH}(s)$ through the modified path. The remainder of $G_{COH}(s)$ is added to $G_{COH}(s)$. This is shown in Figure 7-22 where the constant *Mod* controls the amount that the compensator is modified. When *Mod*=0, the observer is configured as a standard Luenberger observer; when *Mod*=1, the observer is equivalent to the modified structure of Figure 7-19. Any value of *Mod* between 0 and 1 will be a combination of the two, having some portion of the benefits and disadvantages of each. This technique can be used when some filtering is needed but not so much filtering as is provided by the modified observer.

7.5 Exercises

1. Investigate the effects of observer bandwidth on the noise susceptibility of the observed state to sensor noise. Use three sets of tuning gains for the following

exercise: high frequency ($K_{DO}=0.1$, $K_{PO}=45$, $K_{IO}=6400$), medium frequency ($K_{DO}=0.05$, $K_{PO}=17$, $K_{IO}=2000$), and low frequency ($K_{DO}=0.02$, $K_{PO}=8.5$, $K_{IO}=200$).

A. Use Experiment 5A to measure the observer bandwidth of all three sets of gains.

B. Use the DSA in Experiment 7A to measure the noise susceptibility of the observed state to sensor noise with all three sets of gains.

C. Monitor the noise as shown by the *Live Scope* for C_O in Experiment 7A. Set the gains to the low-frequency set. Now, change the gains to the medium-frequency set, one-by-one, monitoring the *Live Scope* display. Repeat for high-frequency gain set. Can you determine the change in noise sensitivity when changing the gains by watching the *Live Scope*? If so, which of the three observer tuning gains has the predominant effect on noise susceptibility?

2. Evaluate noise sensitivity of the power converter output. Recall from Section 7.2.3 that the noise on the power converter output is often a good indicator of audible noise coming from the controller. Use Experiment 7B.

A. Using the PC *Live Scope* display for an approximate measurement, what are the peak-to-peak excursions of the power converter due to noise of the traditional (sensor-feedback) system?

B. Change the feedback *Live Constant* (lower left) to *Observer* to convert the feedback to observer based. Repeat part A with default (high frequency from Exercise 7-1) observer gains.

C. Repeat part A after raising control-loop gains as would normally be done to take advantage of the reduced phase lag of the observer-based system ($K_P=1.5$, $K_I=30$).

D. Using the DSA of Experiment 7B, run Bode plots and evaluate the high-frequency sensitivity of the *PC Output* for all three cases. Compare the results to parts A–C.

3. Compare observer-based disturbance-decoupled and nondecoupled systems using Experiment 7D. Open the file *Experiment_7D.mqd* and save as *Temp.mqd* to avoid permanent changes to the file. Disconnect the *WaveGen* block from the command input and connect it to the unswitched input node of the disturbance DSA (right column, bottom node). At this point, the waveform generator will inject a disturbance into the system. This allows you to monitor the time-domain disturbance response in the *Live Scopes*.

A. Using the *Live Scope*, what is the peak-to-peak excursion of the non-decoupled system when observer feedback is used and control-loop gains are raised accordingly ($K_P=1.5$, $K_I=30$, $K_{DD}=0$, *Feedback = Observer*)?

B. Use disturbance decoupling (set $K_{DD}=1$ and $K_I=0$). What is the excursion?

C. Can you see signs of greater noise susceptibility when disturbance decoupling is enabled?

    D. Reduce the disturbance-decoupling low-pass filter (*DD LPF*) until the excursion is equivalent to the nondecoupled system of part A. What is the value of the filter?

    E. Based on the noise and disturbance DSAs, compare the two systems and comment.

Chapter 8

Using the Luenberger Observer in Motion Control

I n this chapter . . .

- Overview of motion-control and motion-feedback sensors
- Applying observers to improve velocity sensing in encoder and resolver systems
- Applying observers to improve disturbance response in motion systems
- Implementing acceleration feedback based on observed signals

This chapter will discuss several applications of the Luenberger observer in motion-control systems. First, an overview of the operation of motion-control systems is provided. Applications that are likely to benefit from observer technology are presented. The chapter then presents the use of the observer to improve two areas of performance: command response and disturbance response. Command and disturbance response are improved because observed feedback, with its lower phase lag, allows higher loop gains. Disturbance response is further improved through the use of two methods: disturbance decoupling and acceleration feedback. As with earlier chapters, key points will be demonstrated with software experiments.

8.1 The Luenberger Observers in Motion Systems

Motion-control systems are used in a wide range of applications. These applications fall into two broad categories: discrete and continuous operations. In discrete operations, a part is being cut, processed, moved, assembled, or otherwise manufactured. For example, a machine tool may cut metal away from a cast part or a form-fill-and-seal machine might produce a bag of potato chips. Continuous operations run without a definitive end. One example is a machine that produces masking tape, which pulls a roll of paper through a process where adhesive and other coatings are applied.

Motion-control systems are implemented with a number of motor technologies. The lowest cost applications are run *open loop* and do not rely on feedback control loops. For example, stepper motors are usually controlled open loop; they provide an inexpensive way to make discrete moves while controlling position. Variable-speed drives for induction motors, often referred to as *variable frequency AC* or *VFAC* drives, can control motor speed to within a few percentage points without requiring a feed-back device. This technology is often used in applications where position control is not needed, such as when controlling pumps and fans.

The motion-control applications that demand the highest performance are usually implemented with closed-loop *servo* systems. The term *servo* here implies closing velocity and position loops based on physical sensors. Servo systems are among the most complicated motion-control systems to specify and install, and they are usually the most expensive. Engineers select servo systems when the performance of the application demands it. This chapter will focus wholly on servo systems.

8.1.1 Performance Measures

The main measures of performance in servo systems are noise generation, disturbance response, command response, and stability. Servo systems excel in command and disturbance response. Servomotors are often the technology of choice in discrete operations, where the moves must be rapid, and in continuous operations, where the system must hold constant speed in the presence of disturbances.

Noise susceptibility and stability are often problems for servo systems. The high gains needed to respond to rapidly changing commands or high-frequency distur-bances also respond to noise on the sensor and command inputs, often generating considerable noise to the system output. Stability is a problem because designers often push the loop gains as high as possible to maximize response and, in doing so, press the limits of stability margins. As has been discussed in earlier chapters, Luenberger observers are a practical way to deal with stability limitations and so indirectly allow servo gains to be increased. On the other hand, as was the focus of Chapter 7, observers can exacerbate problems with sensor noise.

8.1.2 Servo Control

This section will provide a brief overview of servo control. This material is covered in detail in [11, Chapters 14 and 16]. Servo control here is defined as closed-loop control of velocity or position. In this context, servo control does not include the functions associated with generating torque such as current loops or electronic commutation, the process of channeling current in multiple-winding motors. The assumption here is that the servo loops generate the torque command and other sections of the system control current in order to generate that torque.

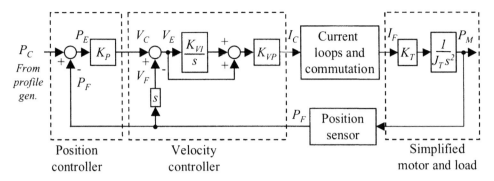

Figure 8-1. Traditional cascaded servo control loops.

Figure 8-1 shows a cascaded servo loop. The position loop is in series or *cascaded* with the velocity loop. In most servo applications the position command, P_C, is generated by a *profile* or *trajectory* generator. The position command is compared to the feedback to generate a position or *following* error. That error is scaled by a position-loop proportional gain, K_P, to create a velocity command, V_C. The velocity loop compares this command to the velocity feedback to create a velocity error. The velocity feedback is commonly created by differentiating the position feedback signal. The velocity control law, which is shown as a PI controller, generates the torque command. For most servomotors, the magnitude of current is approximately proportional to torque and so this signal is commonly used as a current command.

The current signal is fed to the motor control algorithms—current loops and commutation algorithms for brushless motors—which produce motor current in order to generate the torque that will satisfy the loops. The model of the motor here is a simple motor torque constant, which converts current to torque, and a single rotational inertia, J_T.

There are variations of cascaded loops, for example, where the position loop uses a PI law and the velocity loop uses a proportional law. The defining characteristic of the cascaded structure is the use of position error to generate velocity command and the use of velocity error to generate current command. The different variants provide similar performance in similar applications, although the tuning procedures vary considerably.

8.1.2.1 Feed-Forward

Feed-forward is commonly used to improve command response. The command or, more accurately, derivatives of the command are fed ahead of the position loop. This is shown in Figure 8-2 where commanded velocity (V_{PC}) and acceleration (A_{PC}) are fed ahead to the velocity and current loops, respectively. The profile generator can usually provide commanded acceleration and velocity easily since these signals are

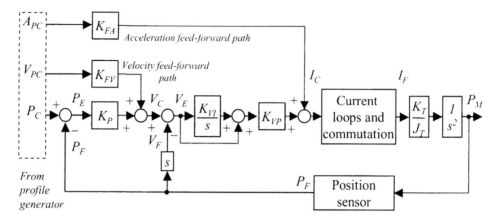

Figure 8-2. Cascaded servo loops with feed-forward.

required to calculate the position command. These signals are fed ahead of the loops so that the inner control loops can respond immediately to the command rather than having to wait for an error signal to percolate through the relatively slow position loop.

Feed-forward is a well-proven technique for improving command response. It does so without increasing the servo gains and thus does not aggravate stability problems or problems from sensor noise. However, the technique is largely unrelated to the most common uses of observers in servo systems and so will not be discussed in detail here.

8.1.2.2 Position vs Velocity Control

Position control is needed for almost all discrete processes. The execution of a discrete process usually requires a mechanism to start and end in known positions. Such motions might be needed to move a silicon wafer, wrap a candy bar, or sew a sequence of stitches. In each case, position control is required to ensure that the sequence begins at the correct position. However, for continuous processes, position control may not be needed. A machine that coats photographic paper usually need not be synchronized to the position of the roll from which the paper comes. Here the servo system need only maintain a constant tension in the coating process, a function that often requires only velocity control. In such cases, the position-loop portion of the cascaded loops shown in Figure 8-1 can be eliminated.

Whether or not a position loop is present, most of the performance issues of a servo system occur in the velocity loop. The velocity loop determines the maximum bandwidth of the control system, even when a position loop is used. This is because the velocity-loop bandwidth is the upper limit of the position-loop gain. High-gain velocity loops are required for high-gain position loops to be stable. So, improving the velocity loop is the primary challenge to those seeking to improve servo-system

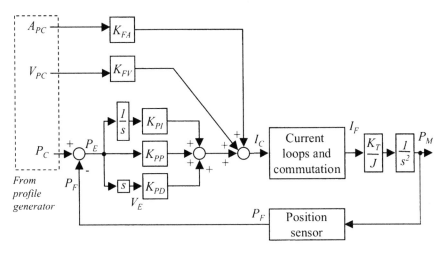

Figure 8-3. PID servo control.

performance. Recognizing this, the discussion and examples of this chapter will focus on the velocity loop.

8.1.2.3 PID Position Loop

A common alternative to the cascaded velocity control loop is the PID position loop. Here there is no explicit velocity loop. Instead, the position error is operated on by a single PID control law. This is shown in Figure 8-3, which includes feed-forward terms.

The PID position loop is quite similar in operation to the cascaded position–velocity loop. For example, the gain K_{PD} in Figure 8-3 operates on velocity error much as the gain K_{VP} does in Figure 8-2. In fact, the high-frequency operation of both loops is almost identical when the two are tuned equivalently. The PID position loop will not be discussed here in detail since its operation is so similar to the cascaded loop. The focus here is the PI velocity loop of Figure 8-2, which is roughly equivalent to the gains K_{PD} and K_{PP} in Figure 8-3.

8.1.3 Common Motion-System Sensors

This section will discuss the most common sensors used to close servo loops. This material is covered in detail in [11, Chapter 13].

8.1.3.1 Tachometer

A tachometer is an electromagnetic device that produces an analog voltage that is proportional to motor speed. Tachometers or *tachs* provide highly resolved, low-phase-lag velocity signals that are ideal for closing velocity loops. In the past, most servo systems relied on tachs for velocity feedback. Unfortunately, the tachometer has numerous shortcomings that have eliminated it from most high-performance motion-control systems.

The primary shortcoming of a tachometer is cost. Since most servo systems must have a position sensor, using a tachometer implies the need for two sensors. The cost of the tachometer itself is substantial, as is the cost of mounting the unit and cabling to it. A second problem with tachometers is brush wear. Tachometers are essentially DC generators, which require carbon brushes to carry voltage between the stator and the rotor. Over time these brushes wear and must be replaced.

Analog tachs are commonly used in low-cost analog servo systems. Their cost is offset by allowing the use of simple, low-cost analog controllers. A small brush motor and a simple analog drive based on tach feedback often cost much less than a digitally controlled brushless-DC servo system.

Tachs are also occasionally used for high-performance servo systems, especially for very low-speed applications. Because a tach is an analog device, it has no explicit resolution limitations in the way that a digital feedback device does. When tachs are wound for high sensitivity (i.e., high-voltage output at relatively low speeds), they can provide high-quality speed signals for low-speed systems.

Tachs are applied at both ends of the servo-performance spectrum—in the lowest cost analog servo systems and high-end systems where precision tachs are used to provide high-resolution velocity feed back. However, most applications rely on a single position sensor from which velocity is derived. Here, the position sensor has dominated as it can feed back a position signal and, indirectly, a velocity signal. The most popular position sensors are encoders and resolvers.

8.1.3.2 Encoders

The optical incremental encoder is probably the most popular position feedback device in modern servo systems. Optical encoders affix an opaque mask to the motor rotor; as the rotor changes position, the pattern of light passing through the mask changes. The encoder produces electronic pulses as the light intensity varies up and down through a certain set point. The optical mask is designed to produce a cyclic pattern of variation that repeats hundreds or thousands of times for each revolution of the motor. The result is that the incremental encoder produces typically between 250 and 5000 counts of position information for each motor revolution. These pulses are produced in proportion to the distance traveled by the encoder. The pulses are counted in the control system to determine how far the encoder rotor has moved.

8.1.3.3 Resolver

Resolvers are electromagnetic devices that are used to sense position. Resolvers, as shown in Figure 8-4, are multiwinding transformers in which the transformation ratio varies with the position of the rotor. A reference winding, *REF*, transmits power on

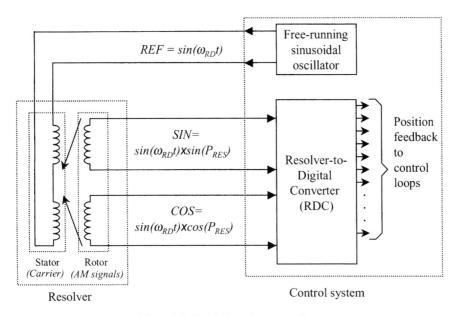

Figure 8-4. Standard resolver connections.

a carrier wave of typically about 5 kHz to the resolver rotor. The two return windings, *SIN* and *COS*, are coupled to the rotor winding where the magnitude of coupling varies with the position of the rotor. The coupling between rotor and *SIN* winding varies according to the sine of the resolver position; the coupling to the *COS* winding varies with the cosine. The *SIN* and *COS* windings are then amplitude-modulated (AM) signals encoding the position of the rotor on the reference carrier.

The conversion of resolver feedback windings to a position signal is more complex than the conversion of incremental encoders. Incremental encoders require only that pulses be counted over time. Resolvers require a scheme that can remove the carrier and provide the necessary trigonometry to produce a measured position. The most common scheme used today is probably the *tracking* or *double-integrating* converter, shown in Figure 8-5. This is a depiction of a typical single integrated-chip (IC) resolver-to-digital (R-D) converter where the conversion process is accomplished with digital and analog components [1, 8].

The double-integrating converter is a closed-loop control system in itself. First, the *SIN* and *COS* signals are demodulated to remove the carrier. Simultaneously, an up–down counter keeps track of the converted position, P_{RD}. The demodulated *SIN* and *COS* signals represent the sine and cosine of the actual position, P_{RES}. These signals are multiplied by the cosine and sine, respectively, of the converted position (P_{RD}) to form $\cos(P_{RES}) \times \sin(P_{RD}) - \sin(P_{RES}) \times \cos(P_{RD})$. By the trigonometry identity $\sin(A - B) = \cos(B) \times \sin(A) - \sin(B) \times \cos(A)$, this can be seen to equal

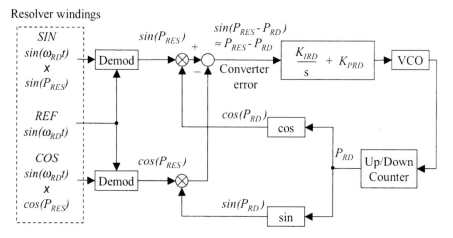

Figure 8-5. Hardware R-D conversion.

$\sin(P_{RES} - P_{RD})$. Assuming the converter error, $P_{RES} - P_{RD}$, is relatively small, $\sin(P_{RES} - P_{RD}) \approx P_{RES} - P_{RD}$ and thus represents the converter error.

After the error signal is derived, it is compensated with a PI control law, shown in Figure 8-5 as $K_{IRD}/s + K_{PRD}$. That error is fed to a voltage-controlled oscillator or VCO, which converts the error to a pulse train. VCOs produce pulses at a frequency in proportion to the magnitude of an analog input. The pulses are counted in the up–down counter, which stores the converted position, P_{RD}.

The dynamic response of the R-D converter can be derived from Figure 8-6, which is a simplified diagram of Figure 8-5. Here, the demodulation, trigonometry, VCO, and up–down counter have been reduced to their equivalent transfer functions. The VCO and up–down counter behave like an integrator. The demodulation and trigonometry act together to produce the error signal, $P_{RES} - P_{RD}$, with few other effects on the dynamic response.

Using the $G/(1 + GH)$ rule in Figure 8-6, the transfer function of the R-D converter is:

$$G_{RD}(s) = \frac{K_{PRD} \times s + K_{IRD}}{s^2 + K_{PRD} \times s + K_{IRD}}. \tag{8.1}$$

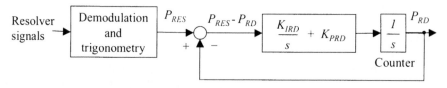

Figure 8-6. Block diagram of R-D conversion.

Upon inspection, this is a low-pass filter. For example, when the frequency is low, the s^2 term, which is in the denominator, is diminished and the transfer function is nearly unity. Also, when the frequency is high, the s^2 term dominates, attenuating the signal and injecting phase lag. The equivalent bandwidth of the process is determined by the gains of the PI control law; these gains are set with passive, discrete components that are connected to the R-D converter as part of a printed circuit board.

The selection of tuning constants generally results in conversion bandwidths of between 300 and 1000 Hz. Processes in the converter induce phase lags in the conversion loop that form the upper bandwidth limit, which is usually 1000 or 1200 Hz, depending on the converter chip. However, in practical systems, noise considerations often force the conversion bandwidth well below that upper limit.

A question often asked about resolvers is why not use an inverse tangent or its equivalent for the conversion to position? This would avoid the complexity of tuning the loop and would remove the phase lag induced by Equation 8.1. There are two key reasons. First, taking an inverse tangent on an integrated circuit such as those used for R-D conversion is more complicated than the approach of Figure 8-5. Second, the noise present on the analog *SIN* and *COS* signals is often substantial; eliminating the double-integrating loop would pass that noise immediately to the position signal. The benefit of the transfer function of Equation 8.1 is that it attenuates high frequencies, albeit at the cost of inducing phase lag.

While the process discussed here is based on hardware R-D converters, a similar process can be carried out in software [22]. Generally, the demodulation remains in hardware, but the other functions are implemented in algorithms executed by a microprocessor or digital signal processor (DSP). The digital converter has important benefits such as reducing parts cost and simplifying adjustment of R-D converter bandwidth. However, the basic principles are the same and the operation of the converters is nearly identical.

8.1.3.4 Sine Encoders

The *sine encoder* has been gaining popularity in servo applications over the past several years. This sensor is a variation on the incremental encoder. Where the incremental encoder transmits digital (two-valued) pulses as position changes, the sine encoder transmits sinusoidal signals that vary continuously as the motor rotates. A typical sine encoder in a servo system produces sine waves with between 512 and 2048 cycles for each motor revolution. Those cycles can be interpolated with a typical resolution of 256 to 1024 positions per sinusoid. So, while an incremental encoder may feed back a few thousand discrete positions per motor revolution, a sine encoder may feed back millions. Sine encoders greatly reduce resolution noise. For reference, a typical resolver has finer resolution than a typical incremental encoder (perhaps 2 to 10 times finer), but both are much coarser than a sine encoder.

One interesting facet of sine encoders is that the signals are often processed in a manner similar to how resolver signals are processed. Because the sine encoder feeds back analog signals that are similar in form to demodulated resolver signals, a tracking loop is often applied to process those signals. As was the case with the resolver, the tracking loop can inject phase lag in the sine encoder signals [6].

8.1.3.5 Deriving Velocity from Position

Most modern servo systems derive velocity information from the position sensor. As stated earlier, this eliminates the need for a separate velocity sensor. The most common method is simple differences: dividing the difference of the two most recently sampled positions by the sample time. Unfortunately, compared to ideal differentiation, using simple differences injects delay equal to half the sample time. This can be seen by evaluating the z-domain transfer function of simple differences ($(z-1)/Tz$, from Table 3-2) and comparing the resulting phase to the ideal 90° advance of true differentiation.

8.1.3.6 Torque Transducers and Tension Control

A minority of servo applications use torque transducers to close servo loops. An example application is high-performance coating machines where the control of the tension of a web line is the most important measure of performance. The term *web* indicates a band of material that is stretched across rollers to be processed. For example, masking tape is produced on web-handling machines. A roll of paper is unrolled onto the web where various coatings are applied; the paper is rewound onto take-up rolls.

It is often critical to maintain a constant tension on the web when applying a coating. Variation in web tension implies a variation in the thickness of the applied coatings. For many processes, such variation lowers the quality of the end product. For example, if the amount of photo emulsion on photographic film varies enough, it can distort the image.

For coating applications, where the variation in coating thickness must be minimized, a torque transducer is sometimes placed on the web. Servomotors control numerous web rolls across which the web is stretched. Because the web is usually constructed of compliant material such as plastics or thin metal foil, tension in the web causes the web material to stretch. For many materials, the stretching approximately obeys Hooke's law, which states that the amount of stretch is proportional to the tension. In that sense, the torque transducer behaves very much like a position sensor. The tension loop, as shown in Figure 8-7 is built as an outer loop to the velocity loop, replacing the position loop of Figure 8-2.

8.1.4 Identifying Applications Most Likely to Benefit from Observers

This section will discuss how to identify motion systems that are most likely to benefit from a Luenberger observer.

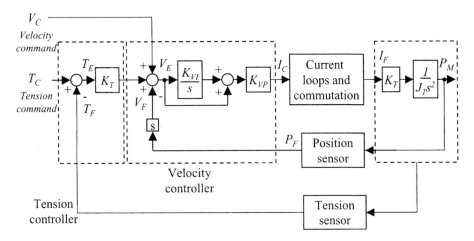

Figure 8-7. Example tension control loop.

8.1.4.1 Performance Requirements

The first area to consider is the performance requirements of the application. Machines that demand rapid response to command changes, stiff response to disturbances, or both will likely benefit from an observer. The observer can reduce phase lag in the servo loop, allowing higher gains, which improve command and disturbance response. Of course, for machines where responsiveness is not an issue, there may be little reason to use an observer.

8.1.4.2 Available Computational Resources

The second factor to consider is the availability of computational resources to implement the observer. Observers almost universally rely on digital control. If the actual or planned control system is executed on a high-speed processor such as a DSP, where computational resources sufficient to execute the observer are likely to be available, an observer can be added without significant cost burden. In addition, if digital control techniques are already employed, the additional design effort to implement an observer is relatively small. However, if the system uses a simple analog controller, putting in place a hardware structure that can support an observer will require a large effort.

8.1.4.3 Controls Expertise in the User Base

Another factor to consider is the user base — the engineers and technicians who purchase, install, and maintain the equipment. Observers require some level of controls expertise for installation and configuration. The user base must be capable of understanding the features of an observer if it is to provide benefit.

8.1.4.4 Sensor Noise

Luenberger observers are most effective when the position sensor produces limited noise. Sensor noise is often a problem in motion-control systems. Noise in servo systems comes from two major sources: EMI generated by power converters and transmitted to the control section of the servo system, and resolution limitations in sensors, especially in the feedback sensor. EMI can be reduced through appropriate wiring practices [30] and through the selection of components that limit noise generation such as those that comply with European CE regulations.

Noise from sensors is difficult to deal with. As was discussed in Chapter 7, Luenberger observers often exacerbate sensor-noise problems. While some authors have described uses of observers to reduce noise, in many cases the observer will have the opposite effect. As discussed in Chapter 7, lowering observer bandwidth will reduce noise susceptibility, but it also reduces the ability of the observer to improve the system. For example, reducing observer bandwidth reduces the accuracy of the observed disturbance signal. The availability of high-resolution feedback sensors raises the likelihood that an observer will substantially improve system performance.

8.1.4.5 Phase Lag in Motion-Control Sensors

The two predominant sensors in motion-control systems are incremental encoders and resolvers. Incremental encoders respond to position change without substantial phase lag. On the other hand, resolver signals are commonly processed with a tracking loop, which generates substantial phase lag in the position signal. Because resolvers produce more phase lag, their presence makes it more likely that an observer will substantially improve system performance. The sine encoder is often processed in a manner similar to the way resolver signals are processed. As was the case with the resolver, the tracking loop can inject substantial phase lag in the sine encoder signals.

Independent of the feedback sensor, most motion-control systems generate phase lag in the control loop when they derive velocity from position. Velocity is commonly derived from position using simple differences. It is well known to inject a phase lag of half the sample time. This phase lag also provides an opportunity for the Luenberger observer to improve system performance.

Five key guidelines for using the Luenberger observer in a motion system are:

- The need for high performance in the application.
- The availability of computational resources in the controller.
- The ability of the average user to install and configure the system.

- The availability of a highly resolved position feedback signal.
- The presence of phase lag in the position or velocity feedback signals.

The first two guidelines are critical — without the need for an observer or a practical way to execute observer algorithms, the observer would not be chosen. The remaining guidelines are important. The more of these guidelines that an application meets, the more likely the observer can substantially improve system performance.

8.2 Observing Velocity to Reduce Phase Lag

The remainder of this chapter will cover ways to use Luenberger observers in motion systems. This section discusses the use of observers to reduce phase lag within the servo loop; as has been discussed throughout this book, removing phase lag allows higher control-law gains, improving command and disturbance response. A common source of phase lag in digital velocity controllers is the use of simple differences to derive velocity from position, as will be discussed in Section 8.2.1. In a resolver-based system, additional phase lag often comes from the method employed to retrieve position information from the resolver, as will be discussed in Section 8.2.2.

8.2.1 Eliminate Phase Lag from Simple Differences

The use of simple differences to derive velocity from position is common in digital motor controllers that rely on a position sensor for feedback. The phase lag induced by the simple differences can be quantified with its transfer function.

$$V_N = \frac{(P_N - P_{N-1})}{T} \tag{8.2}$$

In the z-domain, this becomes

$$V(z) = \frac{P(z) - P(z)/z}{T} \tag{8.3}$$

or

$$\frac{V(z)}{P(z)} = \frac{z-1}{Tz}. \tag{8.4}$$

Equation 8.4 is a z-domain equivalent of differentiation. Ideal differentiation has a phase lead of 90° at all frequencies. Equation 8.4 does produce phase lead, but less than the ideal 90°. The difference between the ideal 90° and simple differences at any given frequency is equivalent to half the sample time, as was discussed in Section 3.2.6.

At any given frequency, f, phase $= T \times f \times 360°$ so that this difference can be expressed as:

$$90° - Phase\left(\frac{z-1}{Tz}\right) = \frac{T}{2} \times f \times 360. \tag{8.5}$$

Equation 8.5 can be confirmed several ways, including by evaluating Equation 8.4 at several frequencies. For example, evaluate Equation 8.4 at $T=0.001$ and $f=100$ Hz. Recall that $z = 1\angle(T \times f \times 360°)$, or here, $z = 1\angle 36°$. Equation 8.4 becomes:

$$\left.\frac{V(z)}{P(z)}\right|_{T=0.001,\, f=100} = \frac{1\angle 36° - 1}{T\angle 36°} = \frac{0.61803\angle 108°}{0.001\angle 36°} = 618.03\angle 72°. \tag{8.6}$$

Notice that Equation 8.6 lags the ideal 90° by 18°, which is, as predicted by Equation 8.5, equal to half the sample time at 100 Hz ($18° = 1000/2\,\mu s \times 100\,Hz \times 360°$). This evaluation can be made at any frequency and will produce the same result.

Incidentally, the magnitude of ideal differentiation is $|s|$ or $2\pi f$. For $f=100\,Hz$, that is equivalent to 628.32 rad/s. Equation 8.6 gives a magnitude of 618.03, which is accurate to within 2%. The magnitude of the simple difference is reasonably accurate, but the phase error is substantial. In fact, for most controls problems, the magnitude of the simple differences is close enough to that of true differentiation that errors in magnitude can be ignored. The phase error between simple differences and ideal differentiation, on the other hand, is substantial at half a sample period and usually must be considered.

8.2.1.1 Form of Observer

The observer structure that eliminates the phase lag generated by simple differences is shown in Figure 8-8. The feedback current, I_F, is scaled by K_T to produce *electromagnetic* torque, T_E, in the physical system, and by K_{TEst} to produce estimated electromagnetic torque, T_{EEst}. The term *electromagnetic* torque describes the torque generated from the windings of the motor. In the actual system, the electromagnetic torque is summed with the disturbance torque, $T_D(s)$, to form the total torque. Total torque is divided by inertia to produce motor acceleration; acceleration is integrated twice to produce velocity and position. The model system is similar except integration, $1/s$, is replaced by its digital equivalent, $Tz/(z-1)$, and the observed disturbance is used in the calculation of observed torque.

The second integration of both the model and the actual system is considered part of the sensor rather than the motor. Because the state of concern is velocity, the assumption in this observer is that the velocity is the plant output, and the integration stage that creates position is an artifact of the measuring system. This observer assumes that the position sensor provides feedback without substantial phase lag. This is a reasonable assumption for an incremental encoder. As has been

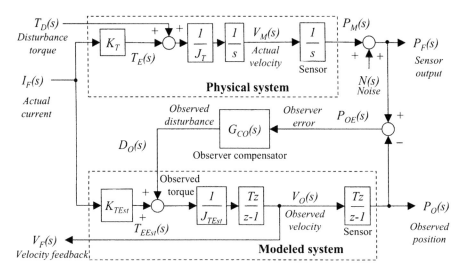

Figure 8-8. Using a Luenberger observer to observe velocity with a fast position sensor.

discussed throughout this book, phase lag incurred when measuring the actual state can be reduced using an observer. In this case, the goal of the observer is to remove the phase lag of simple differences, which would otherwise be used to measure velocity.

8.2.1.2 Experiment 8A: Removal of Phase Lag from Simple Differences

The benefits of the removal of phase lag from simple differences are demonstrated in Experiment 8A. The block diagram of that model is shown in Figure 8-9.[1] This model will be reviewed in detail. The velocity command (*Vc*) comes from a command generator (*Command*) through a dynamic signal analyzer (*Command Dsa*). The command velocity is compared to the feedback velocity and fed to a PI controller (*Digital PI*), which is configured with two *Live Constants*, K_{VP} and K_{VI}. These parameters are set high ($K_{VP}=4.2$, $K_{VI}=850$), yielding a bandwidth of about 660 Hz and a 25% overshoot to a step command; this is responsive performance for the modest 4-kHz sample rate used in this model. As will be shown, the observer is necessary to attain such a high performance level.

The output of the velocity controller is a current command, which feeds the current controller (*Current Loop*). The current controller is modeled as a two-pole low-pass

---

[1]Occasionally, Experiment 8A and some of the other models in Chapter 8 will become unstable. This usually occurs when changing a control-loop gain. In some cases, restoring the model to its original settings will not restore stability. If this happens, click *File, Zero All Stored Outputs*.

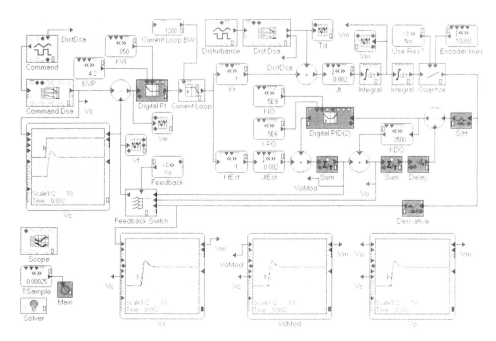

Figure 8-9. Experiment 8A: Investigating the effects of reducing phase lag from simple differences.

filter with a bandwidth of 1200 Hz, set by the constant (*Current Loop BW*), and a fixed damping of 0.7. This is consistent with current-loop controllers used in the motion-control industry.

The output of the current loop is torque-producing current. It feeds the motor through the torque constant (K_T) to create electromagnetic torque. Simultaneously, a disturbance generator (*Disturbance*) feeds a second DSA (*Dist Dsa*) to create disturbance torque (T_D). The two torque signals are summed to create total torque. The total torque is divided by the total inertia (J_T) to produce motor acceleration. (Note that the J_T block is an *Inverse Live Constant*, indicating the output is the input divided by the constant value.) This signal is integrated once to create motor velocity (V_M) and again to create motor position.

There is an optional resolution, which defaults to *not-used*. To enable resolution, double click on *Use Res?* and select *Yes*. This model assumes an instantaneous feedback signal from the position sensor, which is consistent with the operation of encoder systems. Accordingly, the units of resolution are encoder lines, where each line translates into four counts via ×4 *quadrature*, a common method used with optical encoders. The scaling to convert encoder lines to radians (SI units of angle) is implemented with the *Mult* node of the *Live Constant* named *Encoder Lines* being set to $4/2\pi$ or 0.6366, where the 4 accounts for quadrature and the 2π converts revolutions to radians.

The position signal feeds the observer loop, which uses a PID control law (*Digital PID (2)*) configured with two *Live Constants* (K_{PO} and K_{IO}). In the ideal case, the output of the PID controller is equal to acceleration caused by the disturbance torque. The PID output is summed with the power-converter path of the observer, which is the output current scaled by the estimated torque constant (K_{TEst}) and divided by the estimated total inertia (J_{TEst}). The result is summed twice, approximating the two integrals of the physical motor. The derivative gain is configured with K_{DO} in this observer, which is configured as a modified Luenberger observer. The observer produces two output signals: the modified observed velocity, taken just after the first digital integration (the leftmost *Sum*) and the standard observed velocity output, taken just before the second digital integration (the rightmost *Sum*). As in earlier chapters, the delay of one step is added to allow the observer to be constructed since the loop must have some starting point.

One difference between the block diagram of Figure 8-8 and the model of Figure 8-9 is that the observer has been reconfigured to remove the effects of inertia from the observer loop. Notice that J_{TEst} is directly after K_{TEst} and outside the observer loop. This implies that the units of the observer PID output are acceleration, not torque as in Figure 8-8. This will be convenient because changing estimated inertia will not change the tuning of the observer loop; without this change, each time estimated inertia is varied, the observer must be retuned. This observer structure will be used in experiments throughout this chapter.

The model for Experiment 8A includes four *Live Scopes*. At center left is a display of the step response: the command (above) is plotted against the actual motor velocity. Along the bottom are three scopes, from left to right,

(1) the actual motor velocity, V_M, vs the sensed velocity, V_S,
(2) V_M vs the modified observed velocity, V_{OMod}, and
(3) V_M vs the observed velocity, V_O.

The sensed velocity, V_S, lags the motor velocity just a bit because of the phase lag injected by simple differences. As expected, the two observed velocities show no signs of phase lag.

One of the features of the model in Experiment 8A is that it allows the selection of any of the three velocity signals as the feedback signal for the control loop. The switch *Switch4* is controlled with the *Live Constant* named *Feedback*, which can be set to V_S, V_{OMod}, and V_O. Double-click on *Feedback* any time to change the feedback signal. This feature will be used to show the benefit of using the observer to measure velocity.

The key variables used in this and other experiments in this chapter are detailed in Table 8-1. Most are similar to variables in earlier chapters except the general-purpose variables (such as C and R) have been replaced by motion-specific variables (such as V for velocity). Note that all velocity variables display in **RPM** although all calculations are in radians/second (*SI*) units; the scaling is accomplished with *Visual ModelQ* multiplication nodes.

TABLE 8-1 KEY VARIABLES OF EXPERIMENTS 8A–8F

Variable	Description
A_O	Observed motor acceleration
A_{DO}	Observed disturbance, in units of acceleration
Current Loop BW	Bandwidth of current loop which is modeled as a two-pole low-pass filter with a damping ratio of 0.7
I_C, I_F	Command and feedback (actual) current. Input and output of power converter
J_T, J_{TEst}	Actual and estimated total inertia of motor and load. This model assumes a rigid coupling between motor and load
K_{PO}, K_{IO}, K_{DO}	Proportional, integral, and derivative gains of observer
K_T, K_{TEst}	Actual and estimated torque constant of motor. The torque output of the motor windings is the actual current (I_F) multiplied by the torque constant
K_{VP}, K_{VI}	Proportional and integral gains of velocity loop
P_M	Actual motor position
P_F	Feedback motor position
P_O	Observed motor position
P_{OE}	Observed motor position error
T_D	Disturbance torque
V_C	Command velocity
V_E	Velocity error, $V_C - V_F$
V_F	Feedback velocity, the velocity signal used to close the velocity loop, which can be set to V_S, V_O, or V_{OMod} at any time through the constant *Feedback*
V_M	Actual motor velocity
V_O	Observed motor velocity using the standard Luenberger observer
V_{OMod}	Observed motor velocity using the modified Luenberger observer
V_S	Sensed velocity, the velocity derived from the feedback position using simple differences

The results of Experiment 8A are shown in Figure 8-10. There are three plots, one for the system configured using each of V_S, V_O, and V_{OMod} as velocity-loop feedback. The sensed feedback signal has substantial phase lag, which produces ringing in the step response. The two observed velocities produce nearly equivalent results, both having conservative margins of stability. The difference in stability between sensed and observed feedback is due wholly to the phase lag induced by simple differences (Equation 8.5). The two observed feedback signals produce equivalent results as expected; the command responses of the standard and of the modified observers are

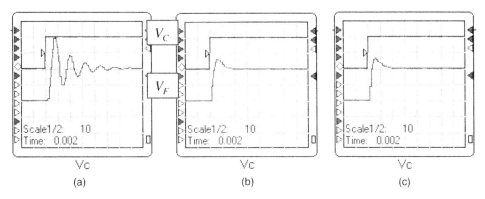

Vc
(a)

Vc
(b)

Vc
(c)

Figure 8-10. Results of Experiment 8A. Square-wave command (above) and response with feedback as (a) sensed velocity (V_S), (b) standard (V_O), and (c) modified (V_{OMod}) Luenberger observer output. Gains are K_{VP}=4.2 and K_{VI}=850 in all cases.

normally equivalent, excepting response to noise and disturbances, neither of which are simulated here. Note that the velocity-loop gains (K_{VP}=4.2, K_{VI}=850) were adjusted to maximize their values when observed velocity is used for feedback; the same values are used in all three cases of Figure 8-10.

The improvement in stability margins can also be seen in Bode plots from Experiment 8A, as seen in Figure 8-11. Here, the use of observed signals (either V_O

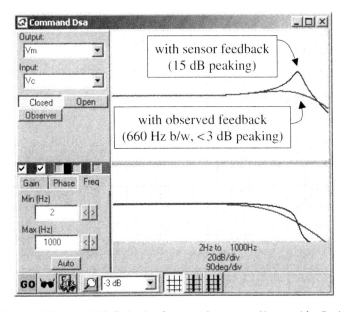

Figure 8-11. Results of Experiment 8A. Bode plot of command response with sensed feedback and observed (V_O) feedback. Gains are K_{VP}=4.2 and K_{VI}=850 in both cases.

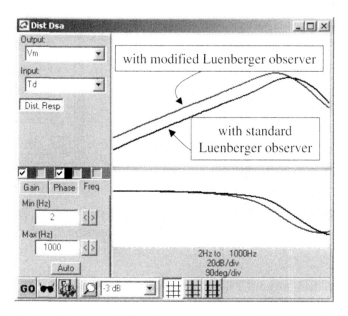

Figure 8-12. Results of Experiment 8A. Disturbance response is degraded for the modified Luenberger observer. Gains are K_{VP}=4.2 and K_{VI}=850 in both cases.

or V_{OMod}) reduces peaking by more than 10 dB, allowing much higher command response than can be supported by sensed feedback. Note that these Bode plots are generated using the *Command Dsa*.

The disturbance response is degraded when using the modified Luenberger observer, as was discussed in Chapter 7. This is demonstrated with the *Dist Dsa* as shown in Figure 8-12. While the command response of both signals is about the same, the disturbance response of the standard form is lower (better) than the modified form below the observer bandwidth. As discussed in Chapter 7, the primary disadvantage of the standard form is increased noise susceptibility, compared to the modified form.

8.2.1.3 Experiment 8B: Tuning the Observer

Experiment 8B isolates the observer to focus on tuning the observer loop. The model is shown in Figure 8-12. The motor and the velocity controller have been removed. The velocity command is summed (i.e., digitally integrated) to produce position command. That command is fed into the observer loop. The observer feed-forward path has been removed so that only the observer loop remains. The model includes a DSA that can show the frequency-domain response of the observer loop. Two *Live Scopes* show the step velocity response of the standard and modified observed velocity. Finally, the variables V_{OE} and V_{O2} are used to display the open-loop response of the observer; these variables will be discussed later in this section.

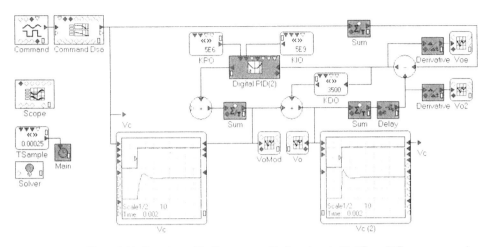

Figure 8-13. Experiment 8B: Observer used in Experiments 8A, 8E, and 8F.

The *Live Scopes* in Figure 8-13 show slight ringing for the selected tuning values ($K_{PO}=5\times10^6$, $K_{IO}=5\times10^9$, $K_{DO}=3500$). These values were determined experimentally, tuning one observer gain at a time. The step response of the standard observer (V_O) for each of the three gains is shown in Figure 8-14. K_{DO} is set just below where overshoot occurs. K_{PO} is set for about 25% overshoot and K_{IO} is set for a small amount of ringing.

The frequency response of the standard observer velocity (V_O) is shown in the Bode plot of Figure 8-15. The bandwidth of the observer is greater than 1000 Hz. In fact, the bandwidth is too great to be measured by the DSA, which samples at 4 kHz. There is 4 dB of peaking, verifying that the observer tuning gains are modestly aggressive.

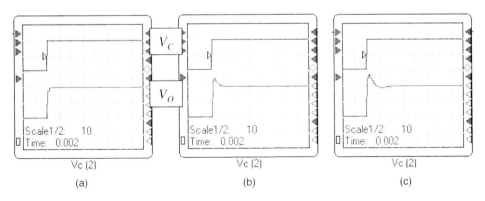

Figure 8-14. Step response of V_O in the tuning of the observer of Experiment 8B and viewing the standard observer output. (a) $K_{DO}=3500$, (b) add $K_{PO}=5\times10^6$, (c) add $K_{IO}=5\times10^9$.

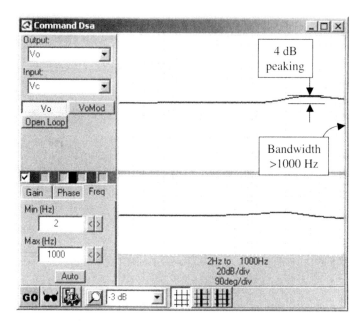

Figure 8-15. Bode plot of observed velocity (V_O) in Experiment 8B.

The frequency response of the modified observer velocity is shown in the Bode plot of Figure 8-16. The bandwidth of the signal is just 500 Hz, well under that of the standard observer. This is consistent with the expected behavior of the modified observer: lower response to both command and noise.

Plotting the open-loop response of the observer loop brings up some interesting modeling issues. First, the loop command, as generated by the waveform generator and DSA, is summed to create the position command to the observer. A detail of this feature, found in Figure 8-13, is shown in Figure 8-17. The implication is that the command is a commanded velocity, which is then integrated to create commanded position. This is required to keep the position command continuous over time. Large step changes in position create enormous swings in velocity and thus can cause saturation in the observer. The integrator is intended to avoid this problem.

A second feature of this model is the use of velocity signals (as opposed to position signals) to provide a Bode plot of the open-loop transfer function of the observer. This is done because the method used to create Bode plots in *Visual ModelQ*, the fast-Fourier transform or FFT, requires the start and end values of each signal being measured to be identical. The position generated from integration of the DSA random number generator does not necessarily meet this requirement. However, the model is configured to start and stop the Bode plot excitation at zero speed, so that the velocities, which are the derivatives of the position signals, always start and stop

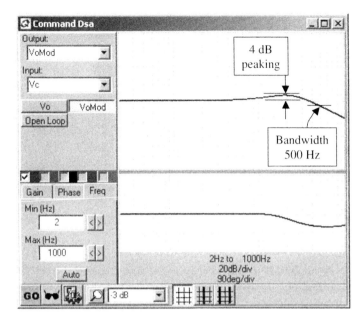

Figure 8-16. Bode plot of modified observed velocity (V_{OMod}) in Experiment 8B.

at zero. Velocity signals can be used in place of position signals because the transfer function of two position signals is the same as the transfer function of their respective velocities. For example, $P_M(s)/P_C(s) = V_M(s)/V_C(s)$. In the s-domain this would be equivalent to multiplying the numerator and denominator by s (s being differentiation), which would produce no net effect. So the open-loop transfer function of the observer, shown in a dashed line in Figure 8-18, is observer position error to observed position; however, that is identical to the derivative of both those signals (shown by the blocks "Derivative"), which produce V_{OE} and V_{O2}. The DSA is preset to show the open loop as V_{O2}/V_{OE}.

The open-loop Bode plot of the observer is shown in Figure 8-19. The gain crossover is at about 700 Hz and the phase margin is about 42°. There are two phase-crossover frequencies, one at about 200 Hz and the other at about 2000 Hz. The gain margin is 12 dB at 200 Hz, but only 6 dB at 200 Hz. The crossover at 200 Hz generates the peaking in Figures 8-15 and 8-16 and the slight ringing in Figure 8-14c.

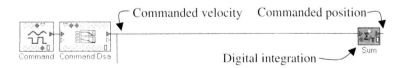

Figure 8-17. Detail of observer command for Experiment 8B (taken from Figure 8-13).

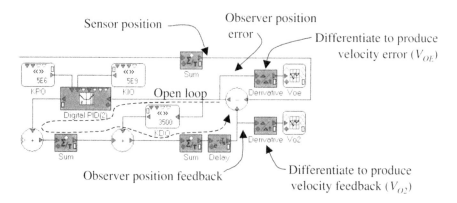

Figure 8-18. Detail of signals for Bode plot from Experiment 8B (taken from Figure 8-13).

The crossover at 200 Hz results because the low-frequency open-loop phase is $-270°$, $-90°$ from each of the two integrators in the observer loop (marked *Sum*) and another $-90°$ from the integral gain in the PID controller, which dominates that block at low frequency. The gain margin at 200 Hz is set almost entirely by the integral gain, K_{IO}. The interested reader can verify this using Experiment 8B (see Exercise 8-1).

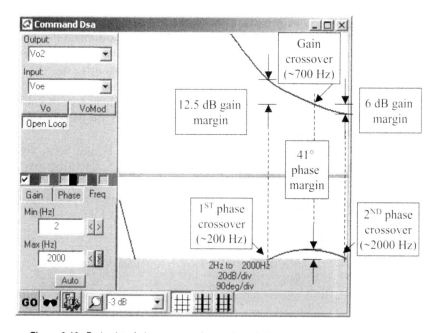

Figure 8-19. Bode plot of observer open-loop gain and phase showing margins of stability.

8.2.2 Eliminate Phase Lag from Conversion

The observer of Figure 8-8 is designed with the assumption that there is not significant phase lag induced in the measurement of position. This is valid for incremental (A-quad-B) encoders, but not for most resolver and sine-encoder conversion methods. Unlike incremental encoders, the conversion of resolver and sine-encoder inputs normally injects significant phase lag. For these sensors, the model needs to be augmented to include the phase lag induced by the conversion of feedback signals. Such an observer is shown in Figure 8-20. Here, the model system sensor includes the R-D converter transfer function of Equation 8.1 as $G_{RDEst}(z)$. Of course, the model of the physical system has also been modified to show the effects of the actual R-D converter on the actual position, $P_M(s)$, using $G_{RD}(s)$.

The observer of Figure 8-20 assumes the presence of a hardware R-D converter. Here, the process of creating the measured position takes place outside of the digital control system. When the process of R-D conversion is done in the digital controller [22], the observer can be incorporated into the R-D converter as is discussed in [14], which is copied in Appendix A.

8.2.2.1 Experiment 8C: Verifying the Reduction of Conversion Delay

The effect of conversion delay is evaluated using Experiment 8C, as shown in Figure 8-21. This experiment uses a two-pole filter as a model of R-D conversion both in the physical system (the block *R-D* at the upper right of the figure) and in the

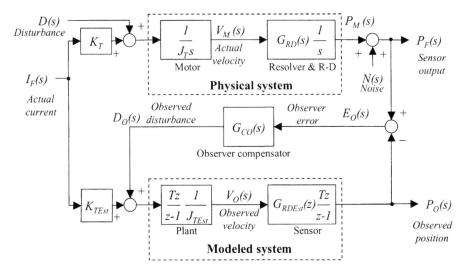

Figure 8-20. Using a Luenberger observer to observe velocity with resolver feedback.

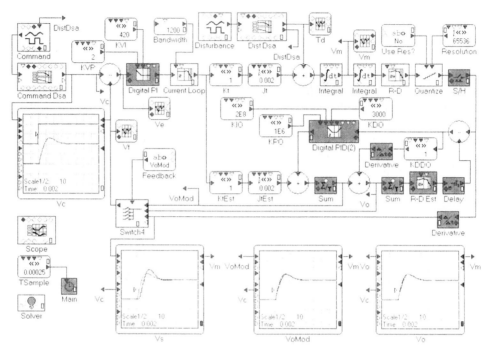

Figure 8-21. Experiment 8C: Reducing phase lag from the R-D converter.

observer model (the block *R-D Est* at the center right of the figure). The filter used to simulate the R-D converter is a low-pass filter with the form:

$$R - D(s) = \frac{2\zeta\omega_N + \omega_N^2}{s^2 + 2\zeta\omega_N + \omega_N^2}.$$

This is an *LPF2A* block in *Visual ModelQ*. This filter has the same form as Equation 8.1. The filter used in the observer, *R-DEst(z)*, is the *z*-domain equivalent of *R-D(s)*.

The observer compensator must change from PID to PID-D^2, where the D^2 indicates a second-derivative term. This second-derivative term is necessary because the *R-D Est* block, being a low-pass filter, adds substantial phase lag. The phase lead added by the derivative term improves stability margins allowing higher observer gains. Without the second derivative, the observer gains are limited to low values as can be verified using Experiment 8D, which will be discussed shortly. In Figure 8-21, the scaling of the second derivative is named K_{DDO}. Note that the *Visual ModelQ* block is a single derivative, but K_{DDO} has the same effect as a second derivative in the observer control law because it is fed in after the first *Sum* block.

As with Experiment 8A (Figure 8-9), Experiment 8C includes four *Live Scopes*. At center left is a display of the step response: the command (above) is plotted against the actual motor velocity. Along the bottom are three scopes, from left to right,

(1) the actual motor velocity, V_M, vs the sensed velocity, V_S,
(2) V_M vs the modified observed velocity, V_{OMod}, and
(3) V_M vs the observed velocity, V_O.

The sensed velocity, V_S, lags the motor velocity by about one-third of a division, much more than the lag seen in Experiment 8A. The phase lag here comes from two sources: simple differences, which Experiment 8A also displayed, and the phase lag from R-D conversion, which is the dominant source of phase lag in this model. As with Experiment 8A, the two observed velocities show virtually no signs of phase lag. Also as with Experiment 8A, Experiment 8C allows the selection of any of the three measured velocities as feedback. Double-click on *Feedback* any time to change the feedback signal.

The benefits of using observed velocity feedback are readily seen by changing the control-loop feedback source via the *Live Constant* named *Feedback*. By default, the source is the modified observed velocity, V_{OMod}. By double-clicking on *Feedback*, the feedback source can be set as the observed velocity, V_O, or as the sensed velocity, V_S, which is the simple difference of the output of the R-D converter. The step response of the system with these three feedback signals is shown in Figure 8-22; in all three cases, the control-loop PI gains are the same: $K_{VP}=2$, $K_{VI}=420$. These values are similar to those used in Experiment 8A except that K_{VP} was reduced to increase stability margins. The results are that the system using V_S is unstable but that the

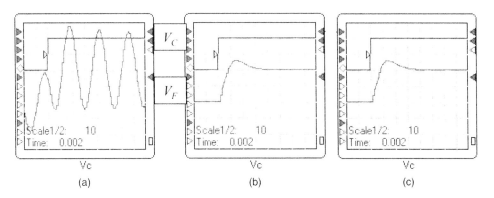

Figure 8-22. From Experiment 8C: Step response with three feedback signals. Both observer feedback types give a similar response, but phase lag in the sensed feedback results in complete instability. For all cases, $K_{VP}=2$ and $K_{VI}=420$. (a) Sensed feedback ($V_F=V_S$); (b) modified observer ($V_F=V_{OMod}$); (c) standard observer ($V_F=V_O$).

two observed signals produce about the same command response; both have reasonable margins of stability. As has been discussed, the differences between using the standard and modified observed feedback are that the system using standard observed velocity will be more responsive to disturbances and more sensitive to sensor noise.

The Bode plot of command response for the gains $K_{VP}=2$ and $K_{VI}=420$ is shown in Figure 8-23. It confirms the results of Figure 8-22. The system based on sensed velocity feedback has over 20 dB of peaking. In fact, the value K_{VP} had to be reduced slightly (from 2.0 to 1.85) to provide sufficient stability margins so that a Bode plot could be calculated (Bode plots cannot be generated from unstable systems). The two cases using observed feedback have a modest 2–3 dB peaking. The dramatic difference in these plots is caused by the ability of the observer to provide a feedback signal that does not include the delays of R-D conversion. In fact, for the observed feedback, the value of K_{VP} can be increased well above 2 while maintaining reasonable margins of stability.

For reference, the R-D converter filters are set for about a 400-Hz bandwidth, which is a typical bandwidth of R-D conversion in industrial systems. Notice that, for the systems based on observed feedback, the closed-loop bandwidth is about 250 Hz. Attaining such high bandwidth with a 400-Hz R-D conversion is very difficult without using observer techniques.

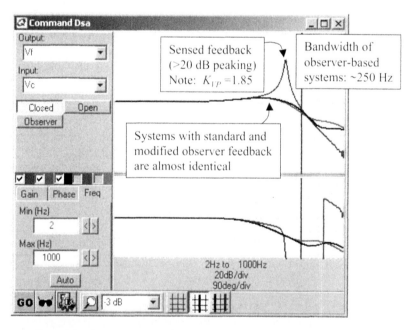

Figure 8-23. From Experiment 8C: Command response with three feedback types ($K_{VP}=2$, $K_{VI}=420$ except K_{VP} was reduced to 1.85 for sensed feedback).

8.2.2.2 Tuning the Observer in the R-D-Based System

The R-D converter for Experiment 8C is tuned to about 400 Hz. In industry, R-D converters are commonly tuned to between 300 and 1000 Hz. The lower the bandwidth, the less susceptible the converter will be to noise; on the other hand, higher bandwidth tuning gains induce less phase lag in the velocity signal. The bandwidth of 400 Hz was chosen as being representative of conversion bandwidths in industry. The response of the model R-D converter is shown in Figure 8-24. The configuration parameters of the filters *R-D* and *R-D Est* were determined by experimentation to achieve 400-Hz bandwidth with modest peaking: *Frequency*=190 and *Damping*=0.7.

The process to tune the observer is similar to the process used in Experiment 8B, with the exception that a second-derivative term, K_{DDO}, is added. Experiment 8D, shown in Figure 8-25, isolates the observer from Experiment 8C, much as was done in Experiment 8B.

The process to tune this observer is similar to that used to tune other observers. First, zero all the observer gains except the highest frequency gain, K_{DDO}. Raise that gain until a small amount of overshoot to a square-wave command is visible. In this case, K_{DDO} is raised to 1, and the step response that results has a small overshoot as shown in Figure 8-26a. Now, raise the next highest frequency gain, K_{DO}, until signs of low stability appear. In this case, K_{DO} was raised a bit higher than 3000 and then backed down to 3000 to remove overshoot. The step response is shown in Figure 8-26b. Next, K_{PO} is raised to 1×10^6; the step response, shown in Figure 8-26c, displays

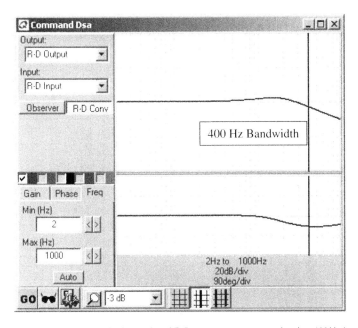

Figure 8-24. From Experiment 8D: Bode plot of R-D converter response showing 400 Hz bandwidth.

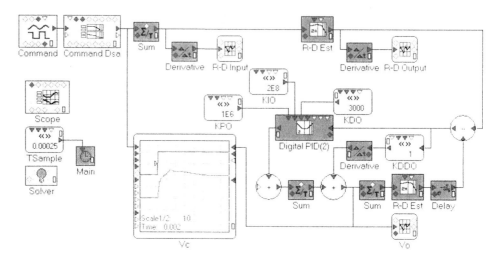

Figure 8-25. Experiment 8D: Tuning the R-D converter.

some overshoot. Finally, K_{IO} is raised to 2×10^8, which generates a slight amount of ringing as shown in the *Live Scope* of Figure 8-25. The Bode plot of the response of the observer gains is shown in Figure 8-27. The bandwidth of the observer is approximately 880 Hz.

▩ 8.3 Using Observers to Improve Disturbance Response

Observers in motion systems provide two distinct ways to improve disturbance response. First, as discussed in Chapter 6, disturbance decoupling can be used to reduce the perturbations caused by disturbances. Second, acceleration feedback can

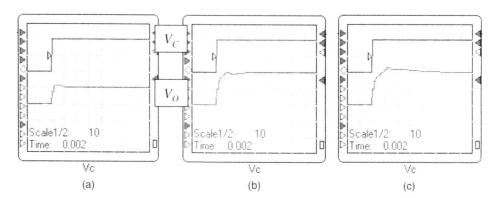

Figure 8-26. From Experiment 8D: Step response in the tuning of the observer. (a) $K_{DDO}=1$, (b) add $K_{DO}=3000$, (c) add $K_{PO}=1 \times 10^6$.

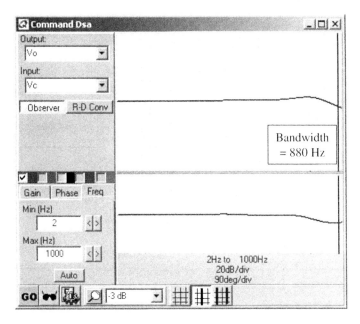

Figure 8-27. From Experiment 8D: Bode plot of response of R-D-based observer tuned for 880 Hz bandwidth $(K_{DDO}=1, K_{DO}=3000, K_{PO}=1\times10^6, K_{IO}=2\times10^8)$.

be used for the same purpose. As will be discussed, while the two methods at first seem unrelated, the structures are similar, as are the benefits.

8.3.1 Disturbance Decoupling in Motion Systems

From Chapter 5, observed disturbance can be used to support disturbance decoupling [25–27, 35, 39]. A configuration that supports disturbance decoupling in motion systems is shown in Figure 8-28. This system is based on a very low phase-lag position sensor, such as an encoder; this can be seen by noticing that the sensor model is the ideal $1/s$ (between V_M and P_M, at upper right). This structure can be adapted for a sensor with significant lag, such as a resolver, by adding the appropriate filter blocks as was discussed in the previous section.

The observer in Figure 8-28 is a standard Luenberger observer in that all three observer-compensation terms are used to form a single signal (the modified observer typically has one path for the derivative term and another for the integral and proportional terms). The output of the observer, A_{DO} in Figure 8-28, is the disturbance torque, but in units of acceleration. This can be seen by noticing that the output flows to two integrators to form position (P_O); a signal that integrates twice to form position must have units of acceleration. Accordingly, the feedback current, I_f, is

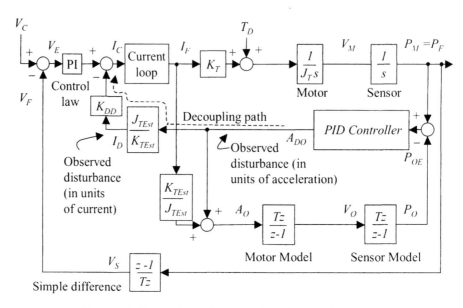

Figure 8-28. Observer-based disturbance decoupling in motion systems.

scaled first by the estimated torque constant, K_{TEst}, to create units of torque and then divided by the total (motor and load) estimated inertia, J_{TEst}, to form units of acceleration. These steps are consistent with the motion-based observers used throughout this chapter.

The unique section of Figure 8-28 is the decoupling path from the observed disturbance, A_{DO}, to the current command. This signal must be converted from units of acceleration to current. This is done by first multiplying by total inertia, J_{TEst}, to create observed torque and then dividing by the motor-estimated torque constant, K_{TEst}, to create current. This signal must also be scaled by the disturbance-decoupling parameter, K_{DD}. If K_{DD} is set to 0, disturbance decoupling is turned off; if it is set to 1, disturbances will be fully decoupled, within the ability of the system.

Recall the stability problems of observer-based disturbance decoupling from Chapter 5. If the observer-loop gain, here affected by J_{TEst} and K_{TEst}, is significantly in error, instability can result. On most servo systems, the value of K_T is known with reasonable precision. However, in many applications, the actual inertia varies considerably during normal operation. Disturbance decoupling is normally difficult to use in applications with varying inertia; however, if the value of J_{TEst} can be regularly updated to follow the changes in actual inertia, the method can still be applied. This normally requires that the changes in inertia are not too rapid and that they can be accurately predicted.

8.3.1.1 Experiment 8E: Using Disturbance Decoupling in Motion Systems

Experiment 8E models the observer structure of Figure 8-28. To simplify the model, a single term K_{TEst}/J_{TEst} is formed at the bottom center of the model (note that the J_{TEst} block is an *Inverse Live Constant*, indicating the output is the input divided by the constant value). This term multiplies the current-loop output to convert that signal from current to acceleration units. The term also divides the observed disturbance to convert its units from acceleration to current. Also, notice that in Figure 8-28 an explicit clamp has been added to the input of the current loop. The clamp function is normally provided implicitly through the PI controller; however, since the PI output is added to the disturbance signal, the output must be clamped again to ensure that the maximum commanded current is always within the ability of the power stage. Another difference between Experiment 8E and the idealized block diagram of Figure 8-28 is that the experiment uses the observed velocity feedback to close the loop in order to get maximum benefit from the observer. The *Live Scope* in Figure 8-29 shows the system response to a torque disturbance (the command source defaults to zero in this model).

The benefits of using disturbance decoupling can be seen by adjusting the parameter K_{DD} to 1 (to fully enable decoupling) as compared to 0 (to disable decoupling). Recall also from Chapter 5 that when K_{DD} is set to 1, the control-loop integral gain needs to be zeroed to avoid ringing in the command step response. The step response

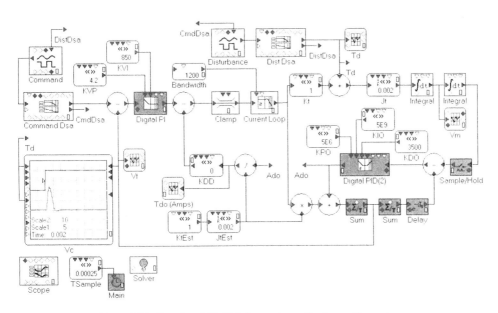

Figure 8-29. Experiment 8E: Disturbance decoupling in a motion system.

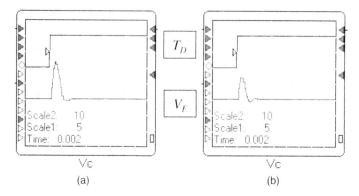

(a)   (b)

Figure 8-30. Response to a 5-nm step torque disturbance without and with disturbance decoupling.
(a) Disabled decoupling ($K_{DD}=0$, $K_I=850$). (b) Full decoupling ($K_{DD}=1$, $K_I=0$).

to a disturbance torque is shown with K_{DD} set to 0 and 1 (with K_I zeroed when $K_{DD}=1$) in Figure 8-30. Disturbance decoupling reduces the excursion in velocity by more than half. Note that the ideal response would be no variation in velocity.

The Bode plot of the disturbance response is shown in Figure 8-31. There are two gain plots, the one above for the system without decoupling and the one below for

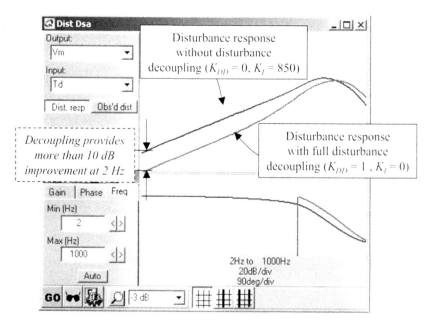

Figure 8-31. Bode plot of disturbance response with and without disturbance decoupling.

the system with full decoupling. These plots are obtained using the *Dist Dsa*, which is at the top center of the model, and setting K_{DD} to both 0 and 1 (and zeroing K_I when $K_{DD}=1$). Notice that the system with disturbance decoupling provides more than 10 dB (10×) enhancement at 2 Hz. This is a substantial improvement, especially taking into account that the velocity-loop gains were already tuned to high values ($K_P=4.2$ and $K_I=850$) in Section 8.2.1.2.

8.3.2 Observer-Based Acceleration Feedback

Acceleration feedback works by slowing the motor in response to measured acceleration [9, 12–14, 21, 25, 27, 32, 38]. The acceleration of the motor is measured, scaled by K_{AFB}, and then used to reduce the acceleration (current) command. The larger the actual acceleration, the more the current command is reduced. K_{AFB} has a similar effect to increasing inertia; this is why acceleration feedback is sometimes called *electronic inertia* or *electronic flywheel* [25, 26]. The idealized structure is shown in Figure 8-32 [12].

The effect of acceleration feedback is easily seen by calculating the transfer function of Figure 8-32. Start by assuming current loop dynamics are ideal: $G_{PC}(s)=1$. Applying the $G/(1+GH)$ rule to Figure 8-32 produces the transfer function

$$\frac{P_M(s)}{I_C(s)} = \frac{K_T/J_T}{1+(K_T/J_T)\times(J_T/K_T)K_{AFB}} \times \frac{1}{s^2} \qquad (8.7)$$

which reduces to

$$\frac{P_M(s)}{I_C(s)} = \frac{K_T/J_T}{1+K_{AFB}} \times \frac{1}{s^2}. \qquad (8.8)$$

It is clear upon inspection of Equation 8.8 that any value of $K_{AFB} > 0$ will have the same effect as increasing the total inertia, J_T, by the factor of $1+K_{AFB}$. Hence, K_{AFB} can be thought of as electronic inertia.

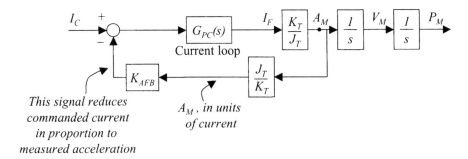

Figure 8-32. Idealized acceleration feedback.

The primary effect of feeding back acceleration is to increase the effective inertia. However, this alone produces little benefit. The increase in effective inertia actually reduces loop gain, reducing system response rates. The benefits of acceleration feedback are realized when control-loop gains are scaled up by the amount that the inertia increases, that is, by the ratio of $1+K_{AFB}$. This is shown in Figure 8-33. Here, as K_{AFB} increases, the effective inertia increases, and the loop gain is fixed so that the stability margins and command response are unchanged.

Using the $G/(1+GH)$ rule and allowing the $1+K_{AFB}$ terms to cancel, the command response for the system of Figure 8-33 is

$$\frac{V_M(s)}{V_C(s)} = \frac{G_C(s) \times G_{PC}(s) \times K_T/(J_T s^2)}{1+G_C(s) \times G_{PC}(s) \times K_T/(J_T s^2)} \tag{8.9}$$

or

$$\frac{V_M(s)}{V_C(s)} = \frac{G_C(s) \times G_{PC}(s) \times K_T/J_T}{s^2 + G_C(s) \times G_{PC}(s) \times K_T/J_T}. \tag{8.10}$$

Notice that the command response is unaffected by the value of K_{AFB}. This is because the loop gain increases in proportion to the inertia, producing no net effect.

The disturbance response of Figure 8-33, unlike the command response, is improved by acceleration feedback. Again, using the $G/(1+GH)$ rule, the disturbance response is:

$$\frac{V_M(s)}{T_D(s)} = \frac{K_T/[(1+K_{AFB})J_T]}{s^2 + G_C(s) \times G_{PC}(s) \times K_T/J_T}. \tag{8.11}$$

For the idealized case of Equation 8.11, the disturbance response is improved through the entire frequency range in proportion to the term $1+K_{AFB}$. For example, if K_{AFB} is set to 10, the disturbance response improves by a factor of 11 across the frequency range. Unfortunately, such a result is impractical. First, the improvement cannot be realized significantly above the bandwidth of the power converter (current loop). This

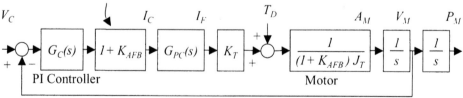

Scale the PI output up by the same amount effective inertia increases.

Figure 8-33. Velocity controller based on idealized acceleration feedback.

is clear upon inspection as the acceleration feedback signal cannot improve the system at frequencies where the current loop cannot inject current. The second limitation on acceleration feedback is the difficulty in measuring acceleration. While there are acceleration sensors that are used in industry, few applications can afford either the increase in cost or the reduction in reliability brought by an addition sensor and its associated wiring. One solution to this problem is to use observed acceleration rather than measured. Of course, the acceleration can be observed only within the capabilities of the observer configuration; this limits the frequency range over which the ideal results of Equation 8.11 can be realized.

8.3.2.1 Using Observed Acceleration

Observed acceleration is a suitable alternative for acceleration feedback in many systems where using a separate acceleration sensor is impractical. Such a system is shown in Figure 8-34. The observed acceleration, A_O, is scaled to current units and deducted from the current command. The term $1 + K_{AFB}$ scales the control-law output as it did in Figure 8-33.

8.3.2.2 Experiment 8F: Using Observed Acceleration Feedback

Experiment 8F models the acceleration-feedback system of Figure 8-34 (see Figure 8-35); the structure is quite similar to the disturbance-decoupling system of Figure 8-29. The velocity loop uses the observed velocity feedback to close the

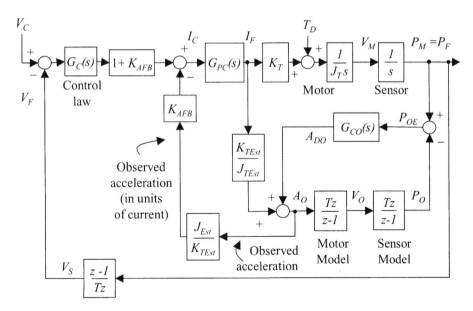

Figure 8-34. Observer-based acceleration feedback in motion systems.

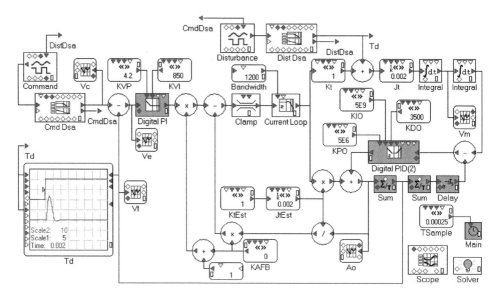

Figure 8-35. Experiment 8F: Using observed acceleration feedback.

loop. Again, a single term, K_{TEst}/J_{TEst}, is formed at the bottom center of the model to convert the current-loop output to acceleration units and to convert the observed acceleration to current units. Unlike Figure 8-35, the term $1+K_{AFB}$ is formed using a summing block (at bottom center) and used to scale the control-law output, consistent with Figure 8-33. As in Figure 8-35, an explicit clamp is used to ensure that the maximum commanded current is always within the ability of the power stage. The *Live Scope* in Figure 8-35 shows the system response to a torque disturbance.

The improvement of acceleration feedback is evident in the step response, as shown in Figure 8-36. As was the case with disturbance decoupling, recall that without acceleration feedback, the control-law gains were set as high as was practical; the non-acceleration-feedback system took full advantage of the reduced phase lag of the observed velocity signal. Still, acceleration feedback produces a substantial benefit.

The Bode plot of the system with and without acceleration feedback is shown in Figure 8-37. The acceleration feedback system provides benefits at all frequencies below about 400 Hz. Figure 8-37 shows two levels of acceleration feedback: $K_{AFB}=1.0$ and $K_{AFB}=10.0$. This demonstrates that as the K_{AFB} increases, disturbance response improves, especially in the lower frequencies where the idealized model of Figure 8-32 is accurate. Note that acceleration feedback tests the stability limits of the observer. Using the observer as configured in Experiment 8F, K_{AFB} cannot be raised above 1.2 without generating instability in the observer. The problem is cured by reducing the sample time of the observer through changing the *Live Constant*

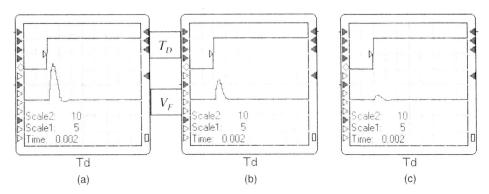

Figure 8-36. Response to a 5-nm step disturbance without and with acceleration feedback. (a) Without acceleration feedback (K_{AFB}=0.0), (b) with minimal acceleration feedback (K_{AFB}=1.0), (c) with more acceleration feedback (K_{AFB}=10.0).

named T_{SAMPLE} to 0.0001. This allows K_{AFB} to be raised to at least 20 without generating instability. It should be pointed out that changing the sample time of a model is easy but may be quite difficult in a working system. See Exercise 8-5 for more discussion of this topic.

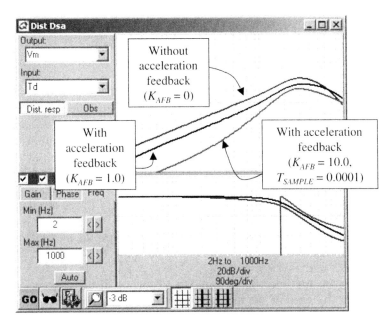

Figure 8-37. Bode plot of velocity loop disturbance response without (K_{AFB}=0) and with (K_{AFB}=1.0, 10.0) acceleration feedback.

8.4 Exercises

1. Verify that at the first phase crossover (200 Hz), the gain margin of the observer in Experiment 8B is set almost entirely by the value of the integral gain, K_{IO}. *Hint: Take the open-loop plot gain with $K_{IO}=0$ and compare to open-loop plot with default K_{IO} at and around 200 Hz.*

2. Use Experiment 8A to compare the disturbance response of observer- and non-observer-based systems with identical control-law gains. Compile and run the model. Change the input from command to disturbance as follows. Double-click on *Command* Wave Gen and set the amplitude to 0.0001 (this is set low enough to be insignificant but high enough to continue to trigger the *Live Scopes*). Similarly, set the amplitude of the *Disturbance* Wave Gen to 2.0. The *Live Scopes* are now showing the velocity signals in response to a disturbance.

   A. Compare the three velocity signals. Notice that the three *Live Scopes* at the bottom compare the three velocity signals to the actual motor velocity.

   B. Predict which signals will give the best disturbance response with the default set of control-loop gains.

   C. Using *Dist DSA*, measure and compare the disturbance response based on each of the three signals.

3. Using the *Live Scope* displays at the bottom of Experiment 8A, compare the noise sensitivity of the three velocity signals. Enable resolution by double clicking on the *Live Constant Use Res?* at top right and select *Yes*. Compare the results here to Exercise 8-2.

4. Use Experiments 8A and 8C to evaluate the robustness of an observer-based system in the presence of fluctuations of total inertia (J_T).

   A. Find the PM of the encoder system of Experiment 8A with nominal parameters, using V_O and V_{OMod} as feedback.

   B. Repeat with the total inertia reduced from 0.002 to 0.0014.

   C. Which system appears more robust, the one dependent on V_O or the one dependent on V_{OMod}?

   D. Repeat part A for the resolver-based system of Experiment 8C.

   E. Repeat part B for the resolver-based system of Experiment 8C.

   F. Repeat part C for the resolver-based system of Experiment 8C.

5. Use Experiment 8F to study the relationship between acceleration feedback and system sample time. Use the *Live Scope* display to see instability caused by excessive values of K_{AFB}.

   A. Make a table that shows the relationship between the maximum allowable K_{AFB} with the system–observer sample time (T_{SAMPLE}) set at 0.00025, 0.0002, and 0.00015.

   B. Compare the improvement in low-frequency disturbance response (say, 10 Hz) for each of the three settings to the baseline system ($K_{AFB}=0$). (Set *TSample* to 0.0001 s for this part; this improves the readings from the DSA without requiring you to change the FFT sample time.)

   C. Compare the results in part B to the ideal improvement gained by using acceleration feedback.

References

1. D.H. Sheingold, Ed., *Analog-Digital Conversion Handbook, 3rd Edition*, Analog Devices, Prentice Hall, Englewood Cliffs, NJ, 1994.
2. A. Antoniou, *Digital Filters Analysis, Design, and Applications, 2nd Ed.* McGraw-Hill, New York, 1993.
3. B. Armstrong-Helouvry, *Control of Machines with Friction*, Kluwer Academic Publishers, New York, 1991.
4. G. Biernson, *Principles of Feedback Controls, Vol. 1.* John Wiley and Sons, New York, 1988.
5. G. Biernson, *Principles of Feedback Controls, Vol. 2.* John Wiley and Sons, New York, 1988.
6. J. Burke, "Extraction of High Resolution Position Information from Sinusoidal Encoders," *Proc. PCIM-Europe*, Nuremberg, (1999): 217–222.
7. J. D'Azzo and C. Houpis, *Linear Control System Analysis and Design 3rd Edition*, McGraw Hill, New York, 1988.
8. Data Devices Corp., Manual for 19220 RDC (19220sds.pdf). Available at www.ddc-web.com (use product search for "19220").
9. J. Deur, "A Comparative Study of Servosystems with Acceleration Feedback," *Proc of IEEE IAS*, Rome, (2000).
10. R. Dorf, *Modern Control Systems (6th Ed.)*, Addison-Wesley, Boston, 1992.
11. G. Ellis, *Control System Design Guide (2nd Ed.)*, Academic Press, Boston, 2000.
12. G. Ellis and R.D. Lorenz, "Resonant Load Control Methods for Industrial Servo Drives," *Proc. of IEEE IAS*, Rome, (2000).
13. G. Ellis, "Cures for Mechanical Resonance in Industrial Servo Systems," *PCIM Proceedings*, Nuremberg, (2001): 187–192. (Copied in Appendix A.)
14. G. Ellis and J.O. Krah, "Observer-Based Resolver Conversion in Industrial Servo Systems," *PCIM Proceedings*, Nuremberg, (2001): 311–316. (Copied in Appendix B.)
15. J. Fassnacht, "Benefits and Limits of Using an Acceleration Sensor in Actively Damping High Frequency Mechanical Oscillations," *Conf. Rec. of the IEEE IAS*, (2001): 2337–2344.

16. G.F. Franklin et al., *Digital Control of Dynamic Systems, 3rd Ed.*, Addison-Wesley, Boston, 1998.
17. G.F. Franklin et al., *Feedback Control of Dynamic Systems, 3rd Ed.*, Addison-Wesley, Boston, 1994.
18. B.K. Ghosh et al., *Control in Robotics and Automation*, Academic Press, Boston, 1999.
19. M. Gopal, *Modern Control System Theory*, John Wiley and Sons, New York, 1993.
20. Y. Hori et al., "Slow Resonance Ratio Control for Vibration Suppression and Disturbance Rejection in Torsional System," *IEEE Trans. Ind. Elec.* **46** (Feb. 1999): 162–168.
21. J.K. Kang and S.K. Sul, "Vertical-Vibration Control of Elevator Using Estimated Car Acceleration Feedback Compensation," *IEEE Trans. on Ind. Elec.* **47** (February 2000): 91–99.
22. J.O. Krah, "Software Resolver-to-Digital Converter for High Performance Servo Drives," *Proc. PCIM-Europe*, Nuremberg, (1999): 301–308.
23. H.S. Lee and M. Tomizuka, "Robust Motion Controller Design for High Accuracy Positioning Systems," *IEEE Trans. Ind. Elec.* **43** (February 1996): 48–55.
24. Y.M. Lee et al., "Acceleration Feedback Control Strategy for Improving Riding Quality of Elevator System," *Proc. of IEEE IAS*, Phoenix, (1999): 1375–1379.
25. R.D. Lorenz, "New Drive Control Algorithms (Stae Control, Observers, Self-Sensing, Fuzzy Logic, and Neural Nets)," *Proc. PCIM Conf.*, Las Vegas, (September 1996).
26. R.D. Lorenz, "Modern Control of Drives," *COBEP'97*, Belo Horizonte, MG, Brazil, (December 1997): 45–54.
27. R.D. Lorenz, "ME-746. Dynamics of Controlled Systems: A Physical Systems-Based Methodology for Non-Linear, Multivariable, Control System Design," Video Class, University of Wisconsin-Madison, 1998.
28. S.J. Mason, "Feedback Theory: Some Properties of Signal Flow Graphs," *Proc. IRE* **41** (July 1953): 1144–1156.
29. S.J. Mason, "Feedback Theory: Further Properties of Signal Flow Graphs," *Proc. IRE* **44** (July 1956): 920–926.
30. S. McClellan "Electromagnetic Compatibility for SERVOSTAR® S and CD." Available at www.motionvillage.com/training/handbook/cabling/shielding.html.
31. D.B. Miron, *Design of Feedback Control Systems*, Harcourt, San Diego, 1989.
32. M.H. Moatemri et al., "Implementation of a DSP-Based, Acceleration Feedback Robot Controller: Practical Issues and Design Limits," *Conf. Rec. IEEE IAS Annual Mtg.*, (1999): 1425–1430.
33. F. Nekoogar and G. Moriarty, *Digital Control using Digital Signal Processing*, Prentice-Hall, Englewood Cliffs, NJ, 1999.
34. W. Palm, *Modeling, Analysis and Control of Dynamic Systems*, John Wiley and Sons, New York, 1983.
35. J.W. Park et al., "High Performance Speed Control of Permanent Magnet Synchronous Motor with Eccentric Load," *Conf. Rec. IEEE IAS Annual Mtg.*, (2001): 815–820.
36. C.L. Phillips and R.D. Harbor, *Feedback Control Systems, 2nd Edition*, Prentice-Hall, Englewood Cliffs, NJ, 1991.
37. O. Rubin, *The Design of Automatic Control Systems*, Artech House, Norwood, MA, 1986.
38. P.B. Schmidt and R.D. Lorenz, "Design Principles and Implementation of Acceleration Feedback to Improve Performance of DC Drives," *IEEE Trans. on Ind. Appl.* (May/June 1992): 594–599.
39. Y.S. Suh and T.W. Chun, "Speed Control of a PMSM Motor Based on the New Disturbance Observer," *Conf. Rec. of IEEE IAS Annual Mtg.*, (2001): 1319–1326.

40. S.M. Yang and S.J. Ke, "Performance Evaluation of a Velocity Observer for Accurate Velocity Estimation of Servo Motor Drives," *Conf Rec. of IEEE IAS Annual Mtg.*, (1998): 1697.

41. J. Yoshitsugu et al., "Fuzzy Auto-Tuning Scheme based on α-Parameter Ultimate Sensitivity Method for AC Speed Servo System," *Conf Rec. of IEEE IAS Annual Mtg.*, (1998): 1625.

42. G.W. Younkin et al., "Considerations for Low-Inertia AC Drives in Machine Tool Axis Servo Applications," *IEEE Trans. on Ind. Appl.* **27** (March/April 1991): 262–268.

Appendix A

Observer-Based Resolver Conversion in Industrial Servo Systems[1]

George Ellis
Kollmorgen, A Danaher Motion Company

Jens Ohno Krah
Kollmorgen Seidel, Germany

Resolvers are commonly used in industrial servo systems. The conversion of resolver signals to measure position is usually accomplished using a tracking loop which causes phase lag between the actual and the measured positions. This phase lag causes instability in the control loop and ultimately reduces performance of the servo system. Observers are well known to reduce phase lag caused by sensors. Observers in servo systems use a combination of the position signal and the torque producing current to observe the motor speed. Resolver-to-digital converters (RDCs) have a structure similar to observers so that RDCs can be modified to behave like observers. This provides several advantages including providing position and velocity feedback with little or no phase lag and providing estimations of motor acceleration and torque disturbance. Acceleration feedback can be used to reduce problems with mechanical resonance. Torque disturbance feedback can be used to improve the dynamic stiffness of the control system.

[1] Presented at PCIM 2001 Conference, Nuremberg, Germany, June 19–21, 2001.
Copyright ZM Communications. Reprinted by permission.

Introduction

Resolvers are multiwinding transformers in which the transformer ratio varies with position. The signals from the resolver are processed to generate a position signal; this process is commonly called *resolver-to-digital conversion* or RDC. RDC is usually structured in a tracking or *double integrating* loop. This loop acts like a filter, reducing the magnitude of high-frequency noise, but also generating lag between the actual position and the RDC output. Phase lag within a control loop is well known to have harmful effects such as reducing stability margins and forcing servo gains to be lowered; ultimately, this phase lag can reduce machine performance.

The use of observers is known to improve the performance of servo controllers. Observers combine knowledge of the plant operation and feedback signals to derive more knowledge of plant states than can be measured from the feedback device alone. A traditional tracking RDC can be restructured as an observer. By combining knowledge of the operation of the servo system with feedback from the resolver, the observer reduces the phase lag of RDC. In addition, the observer can be used to derive motor acceleration and disturbance torque. Acceleration feedback can be used to reduce problems with mechanical resonance. Torque disturbance feedback can be used to improve the dynamic stiffness of the control system.

Resolvers and Traditional RDC

Resolvers are commonly used as position sensing devices. Also, most modern controllers derive velocity feedback from the position sensor by taking the difference of the two most recent positions. Resolvers used in industry fall into two major categories, housed and frameless, both of which are shown in Figure 1. Housed resolvers have independent bearings and an output shaft. Frameless resolvers are provided in two pieces, a rotor and a stator, which are mounted to the motor. Resolvers have several advantages, the most important of which are low cost, rugged construction, and very high reliability.

(a)   (b)

Figure 1. Example of (a) housed and (b) frameless resolvers.

The electromagnetic interaction between the rotor and the stator provides signals from which position can be derived. Resolvers have three windings: a reference, a sine feedback, and a cosine feedback. The reference is a fixed sinusoidal signal, typically with a magnitude of 4–8 V and a frequency of 4–10 kHz. The resolver behaves like a pair of rotating transformers. The transformation ratio from the reference winding to the two feedback windings varies with the position of the resolver rotor. Assuming a reference of $\sin(2\pi5000t)$, the SIN winding will be $\sin(2\pi5000t)\times\sin(P_{RES})$ where P_{RES} is the resolver-rotor position. Similarly, the COS winding value will be $\sin(2\pi5000t)\times\cos(P_{RES})$. This is shown in Figure 2.

Converting the Signal

The resolver provides signals that must be processed in order to derive motor position. The traditional RDC is a single monolithic chip that implements a variety of digital and analog functions [1,2]. As depicted in the conceptual diagram of Figure 3, the SIN and COS signals are demodulated to create the signals $\sin(P_{RES})$ and $\cos(P_{RES})$. Simultaneously, the estimated position, P_{RD}, which is stored in an up–down counter, is fed to specialized D/A converters which produce the signals $\sin(P_{RD})$ and $\cos(P_{RD})$. These signals are multiplied to produce the signal $\sin(P_{RES})\times\cos(P_{RD})-\sin(P_{RD})\times\cos(P_{RES})=\sin(P_{RES}-P_{RD})$. Assuming that the positions from the RDC and of the resolver are fairly close together, $\sin(P_{RES}-P_{RD})\approx P_{RES}-P_{RD}$. In other words, this signal represents the error between the actual position and the measured position.

A PI compensator is applied to the error signal, $\sin(P_{RES}-P_{RD})$. This is usually an op-amp circuit with gains set by discrete resistors and capacitors. The compensator output, which is an analog signal, is converted to a pulse train through a voltage-controlled oscillator or VCO. The output of the VCO is fed to the up–down counter, which acts like an integrator, summing the VCO pulses over time.

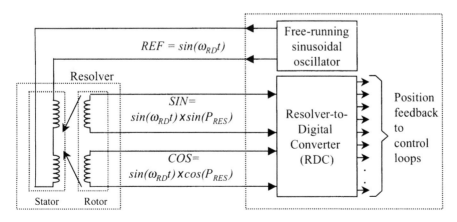

Figure 2. Resolver and R/D converter wiring.

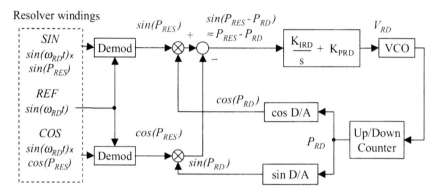

Figure 3. Simplified RDC.

The RDC of Figure 3 is redrawn in Figure 4 to emphasize the effects of the conversion process on the servo system. In Figure 4, the demodulation and trigonometry are combined to create a signal representing the actual position of the resolver, P_{RES}. This signal is compared to RDC position, P_{RD}, to create an error. That error is PI-compensated and fed to an integrator to create P_{RD}. Note that in the traditional RDC, the signal P_{RD} does not explicitly exist. However, dynamics of the RDC are represented accurately in Figure 4.

Figure 4 can be used to derive a transfer function of the RDC using the $G/(1+GH)$ rule [3]. Assuming that demodulation and trigonometric functions do not significantly affect the dynamics (a reasonable assumption), the relationship between the actual resolver position and the output of the RDC is:

$$\frac{P_{RD}(s)}{P_{RES}(s)} = \frac{s \times K_{PRD} + K_{IRD}}{s^2 + s \times K_{PRD} + K_{IRD}}. \tag{1}$$

Equation (1) shows that the RDC behaves like a 2nd order low-pass filter. At low frequencies, where the s^2 denominator term is overwhelmed by the other terms, Equation (1) reduces to one; so, at low frequency, the converter generates no significant effects. However, at high frequencies, the s^2 term will become larger, inducing attenuation and phase lag. Drive manufactures are responsible for selecting and installing the components that set the PI compensator gains. They normally try to maximize

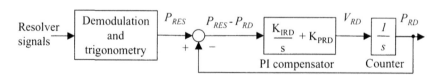

Figure 4. Idealized structure of R/D conversion.

the gains in order to raise the effective bandwidth of the RDC, which minimizes the phase lag induced by the process. However, stability margins and noise combine to limit the bandwidth so the typical RDC has a bandwidth of about 600 Hz.

The main problem caused by induced phase lag of the RDC is the reduction of phase and gain margins in the control system [3]. For high-performance systems, the phase lag of an RDC can cause stability problems such as overshoot and ringing. Ultimately, this can force the reduction of gains, lengthening response times. For example, based on tests carried out by one of the authors, the phase lag induced by RDC limited the velocity-loop bandwidth to about 125 Hz. The removal of the RDC phase lag allowed the bandwidth to be increased to almost 200 Hz while maintaining the same stability margins. For high-performance servo systems, the phase lag of an RDC can be a significant barrier [4].

It should be stated that most of the processes of RDC can also be executed in software [4]. The trigonometry of Figure 3 may be replaced with an inverse tangent function and the VCO and up–down counter can be replaced with a software integrator [5]. Software RDC provides many advantages including reduction in hardware cost and allowing the end user to modify the RDC compensation gains, say, lowering them to attenuate noise. However, the operation of the conversion loop is similar and the dynamics of both are equivalent.

Observers

Observers are commonly used to determine internal states of a system based on measurements of other states. Observers are often applied in cases where the observed states cannot be measured because mounting a sensor is either impractical or too expensive. Observers can also be used simply to improve the quality of measured signals. For example, a resolver measures motor position, but in doing so, it adds considerable phase lag and an observer can remove this lag. This is how the observer will be used here.

The Luenberger observer [6–9], as shown on the right side of Figure 5, observes a state by combining two sources of information: the sensor output ($Y(s)$) and the power converter output ($P_C(s)$). Consider first the path from the power converter (the current controller) which produces a power output that drives the physical system (the motor) which feeds the actual state, $C(s)$; the actual state feeds the actual sensor to produce a sensed output, $Y(s)$. Simultaneously, the power converter output is applied to the model or estimated plant ($GP_{Est}(s)$), which produces an observed state, $C_O(s)$, which feeds a model sensor. The model sensor produces an observed sensor output, $Y_O(s)$. This path is often called the *prediction* path because it predicts where the state will go based on the power applied to the plant.

Ideally, the prediction would be sufficient for producing a high-quality estimate of the actual state. Unfortunately, there are numerous sources of error that degrade that signal. For example, disturbances unknown to the model system affect the actual state. In addition, the plant and sensor models are not exact replicas of their physical

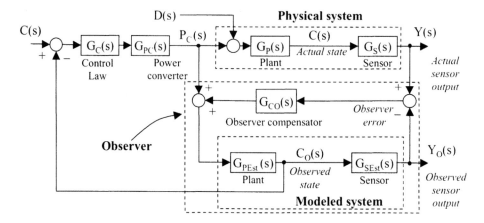

Figure 5. Luenberger observer.

counterparts. These errors corrupt the observed state. The second path in the observer, often called the corrector path, improves the quality of the observed state.

The corrector path runs from the rightmost summing junction in Figure 5 through the observer compensator. That summing junction compares the outputs of the actual and observed sensors to produce observer error. That error is fed into an observer compensator, typically a PID compensator. The compensator output is added to the power converter output before it feeds the model plant. The high-gain observer compensator drives the observer error to near zero so that the actual and observed sensor outputs are nearly identical. If the sensor model is accurate, this forces the observed state to follow the actual state. The observer of Figure 5 produces $C_O(s)$, which is used to close the control loop in place of the sensed output, $Y(s)$.

The benefits of an observer can be understood by considering that the observed state is formed with two signals: the power converter and sensor outputs. A transfer function can be built to show that the observed state relies on the sensor at low frequencies and the power converter output at high frequencies [6]. By adding the power converter output, the observer eliminates the phase lag caused by relying on the sensor. This will be important later when the RDC structure is improved.

Applying the Observer to RDC

The RDC process from Figure 4 is redrawn in Figure 6a. Here, the blocks are rearranged into a style similar to an observer. A factor of $1/s$ is removed from the RDC compensator and moved forward to a separate block. Notice that the RDC compensator appears in the same position as the observer compensator ($G_{CO}(s)$) in Figure 5. Notice also that the first factor of $1/s$ appears in the position of the model plant ($G_{PEst}(s)$) and the second $1/s$ term appears in the position of the model sensor ($G_{SEst}(s)$).

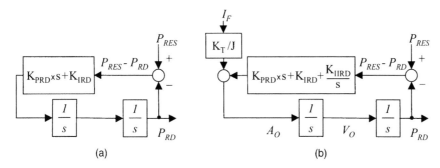

Figure 6. R/D conversion: (a) nonobserver and (b) observer based.

The RDC based on an observer structure is shown in Figure 6b. Here the traditional RDC structure is augmented with the output of the power conversion, I_F. In this structure, the torque-producing current predicts the effects of the motor current on the motor velocity before those effects can be measured with the RDC. This reduces or, in some cases, even eliminates the phase lag caused by the RDC. One other modification is that a third term, K_{IIRD}, is added to the observer compensator. This term is used to remove offset that would otherwise be added to by power converter output.

The output of the observer is affected by the accuracy of the model (here, of K_T/J) and by the tuning gains. The observer must be initialized with an estimate of K_T/J. While moderate errors (for example, 20%) have little effect, large-scale errors will degrade the observed velocity signal significantly. Typically, the magnitude of K_T will be known well enough but inertia often is not known well and, in some applications, even varies significantly during operation. The effectiveness of the observer is reduced in applications with large-magnitude variation of load inertia; in such cases a physical acceleration sensor may be used. The tuning of the observer is similar to tuning for the RDC.

It should be stated that there are a few differences between the traditional observer and the observer shown in Figure 6b. First, this observer moves the motor torque constant and inertia outside the observer loop to have a structure similar to that of the traditional RDC. The plant is simplified to a simple integrator rather than using, say, K_T/Js, a common motor model. Second, the second integrating term on the bottom path is considered part of the sensor, not the plant. This is done since the state of interest is velocity, not position. This is because the dynamics of the velocity loop are so much higher than the position loop that phase lag reduction in the velocity signal is paramount. So, there are some differences between the observer of Figure 6b and the physical system; however, the dynamics of the system are well represented and the velocity signal from the observer is a considerable improvement over the signal from the RDC, as will be demonstrated in the next section.

Advantages of Observer-Based R/D Conversion

The primary advantage of observer-based feedback is the reduction of phase lag in the control loop. Another advantage is the derivation of observed acceleration, which can be used to reduce problems with mechanical resonance. A third advantage is the derivation of observed disturbance torque, which can be used to improve the disturbance response of the drive. In this section, the improvement of phase lag is verified in a lab experiment using Kollmorgen Seidel's ServoStar 600 amplifier. This amplifier formerly relied solely on software-based traditional (nonobserver) RDC [4], but now includes the RDC observer structure.

Reduction in Phase Lag

As discussed in earlier sections, the phase lag created by RDC can cause instability in a servo system, especially when servo gains are raised to high levels. The step response of such a system is shown in Figure 7a. Here the servo gains are set high enough that the system will respond to a 200 RPM step change in speed in 10 ms. However, when the gains are set sufficiently high to produce the necessary responsiveness, the stability margins are reduced and substantial ringing is generated.

The observer improves the stability margins considerably, as shown in Figure 7b. Here, the servo gains are the same, but the ringing is removed. For reference, the servo gains, current loop, and RDC compensation gains are identical. The only difference between Figures 7a and 7b is addition of the power converter path to the RDC.

Derivation of Observed Acceleration Feedback

A second advantage of the observer-based RDC is that observed acceleration feedback is provided through the observer structure. This signal can greatly reduce

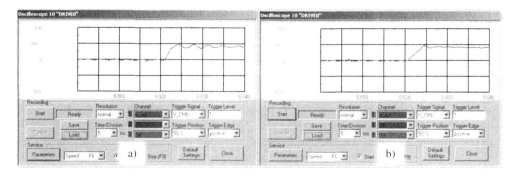

Figure 7. Improvement in stability from reduction of phase lag from (a) a traditional RDC to (b) an observer-based RDC.

problems with mechanical resonance. This subject is discussed in [11], which is also presented at this conference, as well as in [6–10].

Derivation of Observed of Torque Disturbance

A third advantage of the observer-based RDC is that observed torque disturbance is provided by the observer structure. This can be used for disturbance decoupling [3,9], a technique where the disturbance signal is fed back with a polarity inversion to the power converter input. The use of disturbance decoupling can greatly improve the dynamic stiffness of the servo system.

The disturbance decoupled velocity control loop is shown in Figure 8. The output of the observer compensator is marked as T_{DO} or *observed disturbance*. While a detailed discussion of this topic goes beyond the scope of this paper, readers can see that this signal represents the torque disturbance. Notice that the signal is added into the model system after the power converter output and compare this to actual torque disturbance, T_D, which is added to the actual system in approximately the same position. It should be apparent that, assuming the models for the sensor and plant are accurate, the only way for the observer compensator to drive the error to zero is for its output to be equal to the actual torque disturbance. See [9] for more information.

Once the observed torque disturbance signal is available, it can be fed back to the power converter output as shown in Figure 8. So, when a disturbance moves the motor, the observed disturbance signal will provide an estimation of that disturbance, which can be fed to the converter in opposition. Since the observed torque disturbance responds from DC to the observer bandwidth (typically >200 Hz), the response to the disturbance is much more rapid than the response from the velocity loop (typically <100 Hz).

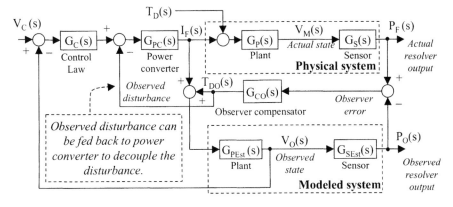

Figure 8. Luenberger observer using disturbance decoupling.

Conclusion

The traditional process of RDC, whether carried out in hardware or software, generates phase lag in servo systems. By restructuring the RDC into an observer, this phase lag can be reduced or even eliminated. The observer uses the power converter output to predict motor movement while the correction path from the sensor ensures that estimation errors do not accumulate. This combination provides a velocity feedback signal that is more representative of actual motor velocity than can be derived from the traditional RDC output. By eliminating phase lag in the velocity feedback signal, the observer allows more responsive servo loops. In addition, the observer provides two other signals: observed motor acceleration and observed torque disturbance. These signals can be employed to further improve servo performance.

References

1. Data Devices Corp, "Manual for 19220 RDC" (19220sds.pdf). Available at www.ddc-web.com (use product search for "19220").
2. D.H. Sheingold, Ed., "Analog-Digital Conversion Handbook" (3rd ed.), Analog Devices, Prentice Hall, Englewood Cliffs, NJ.
3. G. Ellis, "Control System Design Guide" (2nd ed.), Academic Press, Boston, 2000.
4. J.O. Krah, Software resolver-to-digital converter for high performance servo drives, in "Proc. PCIM-Europe 1999, Nuremberg," pp. 301–308.
5. J. Burke, Extraction of high resolution position information from sinusoidal encoders, in "Proc. PCIM-Europe 1999, Nuremberg," pp. 217–222.
6. G. Ellis and R.D. Lorenz, Resonant load control methods for industrial servo drives, in "Proc. of IEEE IAS (Rome), 2000."
7. P.B. Schmidt and R.D. Lorenz, Design principles and implementation of acceleration feedback to improve performance of DC drives, IEEE Trans. Ind. Appl. (May/June 1992), 594–599.
8. M.H. Moatemri, P.B. Schmidt, and R.D. Lorenz, Implementation of a DSP-based, acceleration feedback robot controller: Practical issues and design limits, in "IEEE-IAS Conf. Rec., 1991," pp. 1425–1430.
9. J.K. Kang and S.K. Sul, Vertical-vibration control of elevator using estimated car acceleration feedback compensation, IEEE Trans. Ind. Elec. 47 (2000), 91–99.
10. R.D. Lorenz, Modern control of drives, in "COBEP'97, Belo Horizonte, MG, Brazil, Dec. 2–5, 1997," pp. 45–64.
11. G. Ellis, Cures for mechanical resonance in industrial servo systems, in "PCIM 2001, Nuremberg."

Cures for Mechanical Resonance in Industrial Servo Systems[1]

George Ellis
Kollmorgen, A Danaher Motion Company

Mechanical resonance is a pervasive problem in servo systems. Most problems of resonance are caused by the compliance of power transmission components. Standard servo control laws are structured for rigidly coupled loads. However, in practical machines some compliance is always present; this compliance often reduces stability margins, forcing servo gains down and reducing machine performance.

Mechanical resonance falls into two categories: low frequency and high frequency. High-frequency resonance causes instability at the natural frequency of the mechanical system, typically between 500 and 1200 Hz. Low-frequency resonance occurs at the first phase crossover, typically 200 to 400 Hz. Low-frequency resonance occurs more often in general industrial machines. This distinction, rarely made in the literature, is crucial in determining the most effective means of correction. This paper will present several methods for dealing with low-frequency resonance, all of which are compared with laboratory data.

Introduction

It is well known that servo performance is enhanced when control-law gains are high. However, instability results when a high-gain control law is applied to a compliantly coupled motor and load. Machine designers specify transmission components, such as couplings and gearboxes, to be rigid in an effort to minimize mechanical compli-

---

[1] Presented at PCIM 2001 Conference, Nuremberg, Germany, June 19-21, 2001.

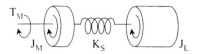

Figure 1. Simple compliantly coupled motor and load.

ance. However, some compliance is unavoidable. In addition, marketplace limitations, such as cost and weight, force designers to choose lighter weight components than would otherwise be desirable. Often, the resulting rigidity of the transmission is so low that instability results when servo gains are raised to levels necessary to achieve desired performance.

The well-known lumped-parameter model for a compliantly coupled motor and load is shown in Figure 1 and a block diagram is shown in Figure 2 [1]. The load and motor are two independent inertias connected by nonrigid components. The equivalent spring constant of the entire transmission is K_S; also, a viscous damping term, b_S, is shown in Figure 2, which produces torque in proportion to the velocity difference of motor and load.

Two-Part Transfer Function

The transfer function from drive current, I_F, to motor velocity, V_M, is

$$\frac{V_M(s)}{I_F(s)} = \left(\frac{K_T}{J_M + J_L} \frac{1}{s}\right)\left(\frac{J_L s^2 + b_S s + K_S}{(J_L J_M (J_L + J_M))s^2 + b_S s + K_S}\right). \tag{1}$$

Equation (1) has two terms. The term on the left is a rigidly coupled motor and load and the term on the right is the effect of the compliant coupling. Note that Equation (1) represents the plant in the case where the feedback sensor is on the motor (as oppose to the load), as is common in industry.

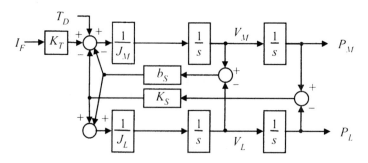

Figure 2. Block diagram of compliantly coupled load.

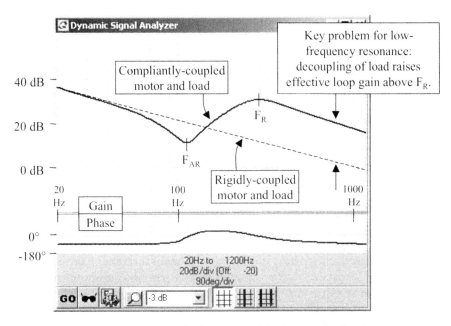

Figure 3. Plot of motor–load plant with 5:1 load-to-motor inertia ratio.

The ideal plant for traditional control laws is a scaled integrator. The term on the left of Equation (1) is such a plant. However, that ideal plant is corrupted by the compliance term (right side of Equation (1)). Figure 3 shows a Bode plot of Equation (1) where the load has about five times more inertia than the motor. The compliance term has a gain peak at the resonant frequency, F_R, and a gain minimum at the antiresonant frequency, F_{AR}, as shown in (2). The corrupting effect of compliance can be seen in Figure 3. Were the load rigidly coupled, the plant would be the ideal integrator (the left side of Equation (1)) which is shown as a dashed line. However, compliance causes attenuation at and around F_{AR} and amplification at, around, and above F_R.

$$F_R = \frac{1}{2\pi}\sqrt{\frac{K_S}{J_P}}\text{Hz}, \quad F_{AR} = \frac{1}{2\pi}\sqrt{\frac{K_S}{J_L}}\text{Hz}. \tag{2}$$

Low-Frequency Resonance

The key problem in low-frequency resonance is the increase in gain at frequencies above F_R [1]. As shown in Figure 3, below F_{AR} the transfer function acts like a simple integrator. The gain falls at 20 dB/decade and the phase is approximately –90°. It also behaves like an integrator above F_R, but with a gain substantially increased compared to the gain well below F_{AR}. Above F_R, the load is effectively disconnected from the motor so that the gain of the plant is the inertia of the motor. In Figure 3, which has a 5:1 load-to-inertia ratio, the effective inertia falls by a factor of 6 dB. This raises the loop gain by 16 dB at high frequencies, reducing margins of stability.

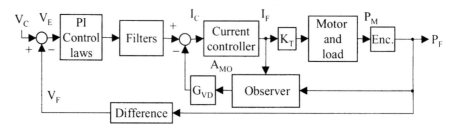

Figure 4. Velocity control system.

Velocity Control Law

Figure 4 shows a velocity control system. The velocity error (V_E) is processed by a control law and filters. The current command, I_C, is connected to the current controller, which produces current, I_F, in the motor. The motor–load plant is connected to an encoder. An observer, fed by the feedback current and position, produces an observed acceleration, A_{MO}. The observer will be discussed in detail later. With exception to the observer, Figure 4 represents a velocity control loop as is commonly used in industry.

The problem caused by the compliant (right) term in Equation (1) is seen in the open-loop Bode plot of the velocity controller of Figure 4. The open-loop transfer function, $V_F(s)/V_E(s)$, is well known to predict stability problems using two measures: phase margin (PM) and gain margin (GM). PM is the difference of −180° and the phase of the open loop at the frequency where the gain is 0 dB. GM is the negative of the gain of the open loop at the frequency where the phase crosses through −180°.

The open-loop plot for a rigidly coupled load demonstrating low-frequency resonance is shown in Figure 5. The harmful effects of compliance are seen in the GM. As marked in Figure 5, when the resonant frequency is well below the first phase crossover (270 Hz) the effect of the compliant load is to reduce the GM approximately by the amount $(J_M + J_L)/J_M$. If J_L/J_M (the *inertia mismatch*) is 5, the reduction of GM will be 6 or about 16 dB. Assuming no other remedy were available, the gain of the compliantly coupled system would have to be reduced by 16 dB, compared to the rigid system, assuming both would maintain the same GM. Such a large reduction in gain would translate to a system with much poorer command and disturbance response.

High-frequency resonance is different; it occurs in lightly damped mechanisms when the natural frequency of the mechanical system (F_R) is well above the first phase crossover. Here, the gain near F_R forms a strong peak. The gain caused by a lightly damped right-hand term in Equation (1) can exceed 40 dB. While both types of resonance are caused by compliance, the relationship of the F_R and the first phase crossover changes the remedy substantially; cures of high-frequency resonance can exacerbate problems with low-frequency resonance. The mechanical structures that cause high-frequency resonance (stiff transmission components and low damping) are

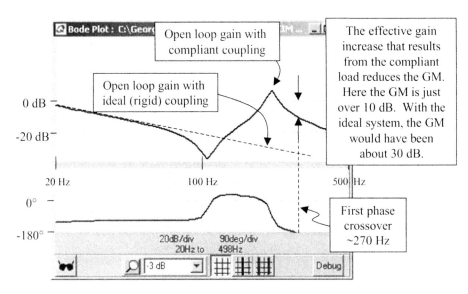

Figure 5. Open loop with low-frequency resonance.

typical of high-end servo machines such as machine tools. Smaller and more cost-sensitive general-purpose machines in such industries as packaging, textiles, plotting, and medical typically have less rigid transmissions and higher damping so that low-frequency resonance is more common.

Methods of Correction Applied to Low-Frequency Resonance

Numerous methods have been used to remedy resonance [1–5]. The most common in industry is the low-pass filter. Another method suggested by numerous authors is acceleration feedback, where the acceleration is provided by an observer. This paper will not discuss the notch filter, which is effective with high-frequency resonance, but not with low-frequency resonance, a problem that appears over a broad frequency range.

Test Unit

The test unit, shown in Figure 6, is a motor and load connected by plastic tubing that has been slit to increase compliance. The motor inertia is 1.8×10^{-5} kg–m^2 and the load is 6.3×10^{-5} kg–m^2 (both include the coupling to the tubing). The coupling has a compliance of 30 Nm/rad. This ratio produced a resonant frequency of about 230 Hz and an antiresonant frequency of about 110 Hz. These figures are consistent with machines used in industry. The drive used was a 3 Amp Kollmorgen ServoStar 600

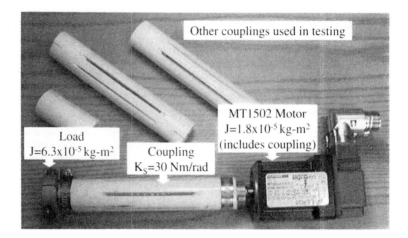

Figure 6. Test-unit mechanism is a scaled model of machines commonly used in industry.

amplifier executing the velocity loop at 16 kHz. This drive is equipped with low-pass and biquadratic filters, an observer, and acceleration feedback.

Baseline System

The baseline system had no filters and did not use acceleration feedback. The PI controller was tuned to maximize performance. The proportional gain was raised as high as possible without generating large oscillations ($G_P=2.5$). The integral gain was then raised until a step command generated 25% overshoot ($G_{ITN}=10$ ms). The step response is shown in Figure 7. Settling time was about 60 ms. The −3-dB bandwidth was measured as 23 Hz.

Low-Pass Filter

A single-pole low-pass filter was used. The frequency was adjusted to 50 Hz. Servo gains could be raised to $G_P=5$ and $G_{ITN}=17$ ms. The settling time improved to 35 ms. The bandwidth was 35 Hz.

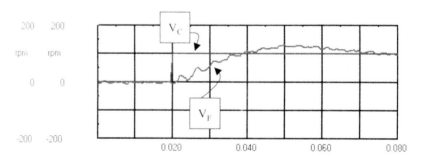

Figure 7. Step response of baseline system.

Biquadratic Filter

A biquadradic filter was applied to the system. A biquad filter has two poles and two zeros; it can be thought of as a high-pass filter in series with a low-pass filter. For these tests, the best case adjustment found experimentally was to set the high-pass filter at 250 Hz and the low-pass filter at 100 Hz. The gains were raised again, this time to $G_V=8$ and $G_{VTN}=10$ ms. Performance improved further. Settling time was reduced to 22 ms. The bandwidth was measured as 56 Hz.

Acceleration Feedback

Acceleration feedback was applied [6–15]. Acceleration was observed using a Luenberger observer, as shown in Figure 8. The observer takes input from the motor current and the encoder. It adds the two and feeds the sum to a model of the motor. That model produces the observed position, which is compared to the actual position. The PID observer compensator drives out most error up to the observer bandwidth, which is usually between 200 and 500 Hz. One by-product of the observer is an acceleration signal, which represents acceleration much better than double-differentiating the position feedback signal. A more complete discussion of the observer can be found in [6,9–11].

Acceleration feedback was applied to the biquad system. The gain G_{VD} was set to 15, which is approximately 2.5 in the SI units shown in Figure 8. This reduced the effects of resonance and the PI gains were raised ($G_V=22$ and $G_{VTN}=12$). The results were a dramatic improvement over the other systems. Settling time was reduced to just 12 ms and the bandwidth was measured as 77 Hz. The step response is shown in Figure 9.

Comparison of All Methods

All three methods (low pass, biquad, and acceleration feedback–biquad) are compared in the frequency domain with the closed-loop Bode plot ($V_F(s)/V_C(s)$) of Figure 10. This plot is generated by the ServoStar 600 Drive on the test mechanism

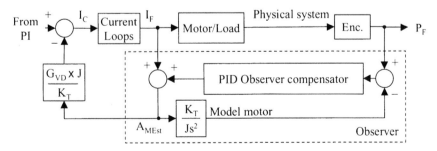

Figure 8. Luenberger observer provides observed acceleration (A_{MEst}).

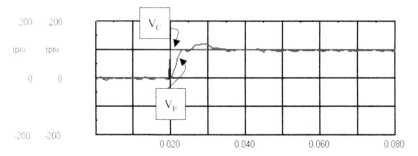

Figure 9. Step response of system with acceleration feedback.

of Figure 6. The results show the dramatic improvement offered by acceleration feedback. Notice that the bandwidth (the frequency where the gain falls to −3 dB) is greatly increased. In addition, stability margins are maintained. Peaking, the undesirable phenomenon where gain rises above 0 dB at high frequency in the closed-loop response, is a reliable measure of stability. As shown in Figure 10, the peaking of all four configurations is about the same, with the baseline system displaying the most peaking. This indicates that the cures for resonance allow higher gain while maintaining equivalent margins of stability.

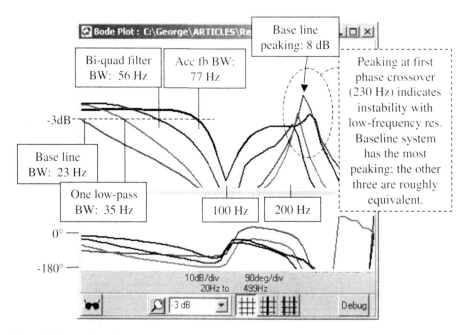

Figure 10. Comparison of closed-loop responses of baseline and when using three antiresonance methods.

Conclusion

Low-frequency resonance is common in industry. It differs from the less common, but more often studied problem of high-frequency resonance. Low-frequency resonance causes oscillations at the first phase crossover of the open-loop system; high-frequency resonance causes oscillations at the natural mechanical frequency of the mechanism. Low-frequency resonance can be addressed by several methods including low pass, biquadratic filters, and acceleration feedback. Acceleration feedback is a practical method for curing low-frequency resonance. It has been implemented in an industrial motor controller.

Acceleration feedback in combination with the biquad filter provides dramatic improvement for systems suffering from low-frequency resonance. This was demonstrated on laboratory hardware. Compared to the traditional solution of a single-pole low-pass filter, the combination of acceleration feedback and the biquad filter allow the settling time to be cut by a factor of three (from 35 to 12 ms) and the bandwidth to be raised by that same factor (23 to 77 Hz). At the same time, acceleration feedback maintained stability margins, indicating that the increased gain of the servo system will be useful in practical applications.

Acknowledgments

The author thanks Dr. Jens Krah for his contributions to this paper, which include implementing the observer and acceleration feedback in the ServoStar 600 and providing numerous helpful insights.

References

1. G. Ellis, "Control System Design Guide" (2nd ed.), Academic Press, Boston, 2000.
2. P. Schmidt and T. Rehm, Notch filter tuning for resonant frequency reduction in dual inertia systems, *in* "Proc. of IEEE IAS, Phoenix, Oct. 3–7, 1999," pp. 1730–1734.
3. S. Vukosavic and M. Stojic, Suppression of torsional oscillations in a high-performance speed servo drive, *IEEE Trans. Ind. Elec.* **45** (1998), 108–117.
4. H. Wertz and F. Schütte, Self-tuning speed control for servo drives with imperfect mechanical load, *in* "Proc of IEEE IAS, Rome," 2000.
5. R. Dhaouadi, K. Kubo, and M. Tobise, Analysis and compensation of speed drive systems with torsional loads, *IEEE Trans. Ind. Appl.* **30** (1994), 760–766.
6. G. Ellis and J.O. Krah, Observer-based resolver conversion in industrial servo systems, *in* "PCIM-Europe, 2001."
7. Y. Hori, H. Sawada, and Y. Chun, Slow resonance ratio control for vibration suppression and disturbance rejection in torsional system, *IEEE Trans. Ind. Elec.* **46** (1999), 162–168.
8. J. Deur, A comparative study of servosystems with acceleration feedback, *in* "Proc of IEEE IAS, Rome, 2000."
9. G. Ellis and R.D. Lorenz, Resonant load control methods for industrial servo drives, *in* "Proc. of IEEE IAS, Rome, 2000."

10. P.B. Schmidt and R.D. Lorenz, Design principles and implementation of acceleration feedback to improve performance of DC drives, *IEEE Trans. Ind. Appl.* (May/June 1992), 594–599.

11. M.H. Moatemri, P.B. Schmidt, and R.D. Lorenz, Implementation of a DSP-based, acceleration feedback robot controller: Practical issues and design limits, *in* "IEEE-IAS Conf. Rec., 1991," pp. 1425–1430.

12. Y.M. Lee, J.K. Kang, and S.K. Sul, Acceleration feedback control strategy for improving riding quality of elevator system, *in* "Proc. of IEEE IAS, Phoenix, 1999," pp. 1375–1379.

13. J.K. Kang and S.K. Sul, Vertical-vibration control of elevator using estimated car acceleration feedback compensation, *IEEE Trans. Ind. Elec.* **47** (2000), 91–99.

14. R.H. Welch, "Mechanical Resonance in a Closed-Loop Servo System: Problems and Solutions," Tutorial from Welch Enterprises, Oakdale, MN.

15. G.W. Younkin, W.D. McGlasson, and R.D. Lorenz, Considerations for low-inertia AC drives in machine tool axis servo applications, *IEEE Trans. Ind. Appl.* **27** (1991), 262–268.

Appendix C

European Symbols for Block Diagrams

This appendix lists the symbols for the most common function blocks in formats typically used in North America and in Europe. Block diagram symbols in most North American papers, articles, and product documentation rely on text. Most linear functions are described by their s-domain or z-domain transfer functions; nonlinear functions are described by their names (*friction* and *sin*). On the other hand, block diagram symbols in European literature are generally based on graphical symbols. Linear functions are represented by the step response. There are exceptions in both cases. In North American documentation, saturation and hysteresis are typically shown symbolically, whereas European literature uses text for transcendental functions.

Part I: Linear Functions

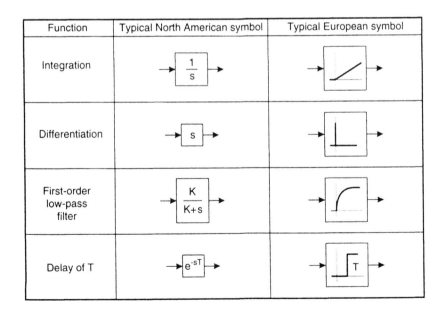

Function	Typical North American symbol	Typical European symbol
Integration	$\dfrac{1}{s}$	
Differentiation	s	
First-order low-pass filter	$\dfrac{K}{K+s}$	
Delay of T	e^{-sT}	

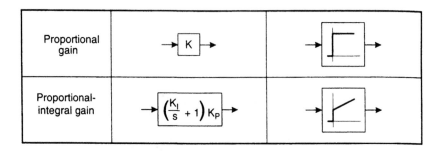

Proportional gain	$\rightarrow$ K $\rightarrow$	
Proportional-integral gain	$\rightarrow \left(\dfrac{K_I}{s} + 1\right) K_P \rightarrow$	

Part II: Nonlinear Functions

Function	Typical North American symbol	Typical European symbol
Rectification		
3-Phase Rectification		
AC Inverter		
D-to-A and A-to-D	D-to-A A-to-D	
Resolver-to-digital converter	R-to-D	
Hysteresis		

Saturation (clamping)	*Clamp*	
Saturation (clamping) with synchronization (anti wind-up)	*Anti wind-up* $\left(\dfrac{K_I}{s} + 1\right) K_P$	
Inductance with saturation	$v \rightarrow \dfrac{1}{(L_0 + L_2 i^2)s} \rightarrow i$	
Transcendental functions (e.g., sin)	$\rightarrow$ sin $\rightarrow$	$\rightarrow$ sin $\rightarrow$
Voltage-controlled oscillator (VCO)	$\rightarrow$ VCO $\rightarrow$	$\rightarrow$ VCO $\rightarrow$
Friction	$\rightarrow$ Friction $\rightarrow$	

Appendix D

Development of the Bilinear Transformation

Bilinear Transformation

The bilinear transformation is so named because it approximates z with a ratio of two linear functions in s. Begin with the definition of z:

$$z \equiv e^{sT} = \frac{e^{sT/2}}{e^{-sT/2}} \tag{D.1}$$

The Taylor series for e^{sT} is

$$z = 1 + sT + \frac{(sT)^2}{2!} + \frac{(sT)^3}{3!} + \ldots \tag{D.2}$$

Using the first two terms of the Taylor series for both the numerator and denominator of Equation D.1 produces

$$z \cong \frac{1 + sT/2}{1 - sT/2} \tag{D.3}$$

Some algebra rearranges the equation to

$$z \cong \frac{2}{T}\left(\frac{z-1}{z+1}\right) \tag{D.4}$$

As an alternative to Table 3-2, Equation D.4 can be used to provide a transfer function in z that approximates any function of s.

Prewarping

Prewarping the bilinear transformation causes the phase and gain of the s-domain and z-domain functions to be identical at the prewarping frequency. This is useful where exact equivalence is desired at a particular frequency, such as when using a notch filter. Prewarping modifies the approximation of Equation D.4 to:

$$s \approx \frac{\omega_0}{\tan(\omega_0 T/2)}\left(\frac{z-1}{z+1}\right) \tag{D.5}$$

where ω_0 is the prewarping frequency, the frequency at which exact equivalence is desired. Recalling Euler's formulas for sine and cosine:

$$\cos(x) = \frac{e^{jx} + e^{-jx}}{2}, \quad \sin(x) = \frac{e^{jx} - e^{-jx}}{2j} \tag{D.6}$$

and recalling that $\tan(x) = \sin(x)/\cos(x)$, Equation D.5 can be rewritten as

$$s = \omega_0 \cdot \frac{2j}{e^{j\omega_0 T/2} - e^{-j\omega_0 T/2}} \cdot \frac{e^{j\omega_0 T/2} + e^{-j\omega_0 T/2}}{2}\left(\frac{z-1}{z+1}\right)$$

$$= j\omega_0 \cdot \frac{e^{j\omega_0 T/2} + e^{-j\omega_0 T/2}}{e^{j\omega_0 T/2} - e^{-j\omega_0 T/2}}\left(\frac{z-1}{z+1}\right)$$

$$= j\omega_0\left(\frac{e^{j\omega_0 T/2} + e^{-j\omega_0 T/2}}{e^{j\omega_0 T/2} - e^{-j\omega_0 T/2}}\right)\left(\frac{e^{sT} - 1}{e^{sT} + 1}\right) \tag{D.7}$$

Our interest here is in steady-state response so that $s = j\omega$:

$$s = j\omega_0\left(\frac{e^{j\omega_0 T/2} + e^{-j\omega_0 T/2}}{e^{j\omega_0 T/2} - e^{-j\omega_0 T/2}}\right)\left(\frac{e^{sT} - 1}{e^{sT} + 1}\right) \tag{D.8}$$

Now, if $e^{j\omega T/2}$ is divided out of both the numerator and the denominator (on the right side), the result is

$$s = j\omega_0\left(\frac{e^{j\omega_0 T/2} + e^{-j\omega_0 T/2}}{e^{j\omega_0 T/2} - e^{-j\omega_0 T/2}}\right)\left(\frac{e^{j\omega T/2} - e^{-j\omega T/2}}{e^{j\omega T/2} + e^{-j\omega T/2}}\right) \tag{D.9}$$

So when $\omega = \omega_0$, most of the factors cancel out, leaving the exact value for s:

$$s = j\omega$$

which means that when the transfer function is evaluated at the prewarping frequency, the approximation is exactly correct.

Factoring Polynomials

Most methods of approximating functions of s with functions of z require that the polynomials, at least the denominator, be factored. The bilinear transformation does not have this requirement, though the factored form usually requires less algebra, as this example shows. Compare this function factored (D.10) and unfactored (D.11):

$$T(s) = \frac{1}{(s+1)^4} \qquad (D.10)$$

$$T(s) = \frac{1}{s^4 + 4s^3 + 6s^2 + 4s + 1} \qquad (D.11)$$

The factored form can be converted to z almost directly:

$$\frac{[T \times (z+1)]^4}{[T(z-1) + T(z+1)]^4} = \frac{[T/(T+2)]^4 (z+1)^4}{[z + (T-2)/(T+2)]^4} \qquad (D.12)$$

However, the unfactored form would require a considerable amount of algebra to convert.

Phase Advancing

The approximation, $z + 1 \approx 2\angle\omega T/2$, can be used to advance the phase of the z function when the s function has fewer zeros than poles.

To begin:

$$z + 1 = e^{j\omega T} + 1 \qquad (D.13)$$

Dividing $e^{j\omega T/2}$ out of the right side yields

$$z + 1 = e^{j\omega T/2}(e^{j\omega T/2} + e^{j\omega T/2}) \qquad (D.14)$$

Recalling Euler's formula, $2\cos(x) = e^{jx} + e^{-jx}$, produces

$$z + 1 = 2e^{j\omega T/2}\cos\left(\frac{\omega T}{2}\right) \qquad (D.15)$$

Finally, when $\omega T/2$ is small, $\cos(\omega T/2) \approx 1$ so that

$$z + 1 \approx 2e^{j\omega T / 2} = 2 \angle \left(\frac{\omega T}{2} \right) \tag{D.16}$$

This approximation is accurate enough for most applications since it is usually not important that the gain of the s and z functions match at high frequencies.

Appendix E

Solutions to Exercises

Chapter 2

2-1A. Compile, click run, and double-click on the block K_P to bring into view an adjustment box. Click on bold >> button to raise value. $K_P=2$ induces ringing and overshoot; $K_P=5$ causes instability. Conclusion: Higher control-loop gains reduce stability margins.

2-1B. Double-click on *Wave Gen* block to bring the *Wave Gen* control panel into view. Use combo box at left to change waveform type. Conclusion: Gentler excitation makes marginal stability more difficult to recognize.

2-1C. Overshoot is gone with zero integral gain; overshoot increases with greater K_I. Conclusion: Higher integral gain causes overshoot and instability.

2-2A. 176 Hz.

2-2B. −11.4 dB.

2-2C. Lower control-loop gains reduce command response at higher frequencies.

2-3A. 0.002 to 0.0002 s.

2-3B. More stable, but only over a range. Below 0.0002 s, little improvement can be seen. This is because the phase lag of the power converter and feedback filter are large enough to make insignificant the phase lag associated with sampling.

2-3C. The two are almost the same (notice all model parameters are identical). Conclusion: A digital PI system with sufficiently fast sample time behaves approximately like an analog PI system.

Chapter 3

3-1A. $K_P \approx 0.5$, $K_I \approx 42$. GM = 13 dB, PM = 60°.

3-1B. $K_P \approx 0.85$, $K_I \approx 70$. GM = 12 dB, PM = 58°.

3-1C. $K_P \approx 1.0$, $K_I \approx 85$. GM = 12.5 dB, PM = 59°.

3-1D. $K_P \approx 1.4$, $K_I \approx 120$. GM$=12.8$ dB, PM$=59.5°$
3-1E.   No overshoot with K_P and 10% with K_I gives 12–13 dB GM and 58°–60° PM.
3-1F.   Allow some overshoot with K_P and/or more than 10% with K_I.

3-2A.   76, 140, 157, and 214 Hz.
3-2B.   If tuning criteria are held constant, more phase lag from filters will reduce the ultimate responsiveness of a control system.

3-3A.   Procedure (determined by trial-and-error): Allow about 5% overshoot with K_P and about 25% overshoot with K_I.
       $K_P \approx 0.7$, $K_I \approx 70$. GM$=10.3$ dB, PM$=49°$.
3-3B. $K_P \approx 1.0$, $K_I \approx 120$. GM$=10.3$ dB, PM$=49°$.
3-3C. $K_P \approx 1.4$, $K_I \approx 140$. GM$=9.5$ dB, PM$=48°$.
3-3D. $K_P \approx 2.0$, $K_I \approx 200$. GM$=9.5$ dB, PM$=49°$

3-4A.   124, 180, 255, and 363 Hz.
3-4B.   The bandwidths in 3-4A are about 50% higher than those in 3-2A. More aggressive tuning leads to higher frequency command response.

Chapter 4

4-1A. $K_P = 0.6$, $K_I = 10.0$.
4-1B.   Approximately 0.25 s.
4-1C. $K_P = 1.7$, $K_I = 30.0$. Settles in approximately 0.1 s.
4-1D.   Luenberger observer removes phase lag. This allows higher control-law gains which provide faster settling time.

4-2A. $K_{DO} = 0.05$, $K_{PO} = 17$, $K_{IO} = 2000$.
4-2B.   108 Hz.
4-2C.   No significant difference. Tuning values do not depend on observer bandwidth, at least when the observer models are very accurate.

4-3A.   Yes, nearly identical.
4-3B.   No. Phase lag can cause instability in the loop, even when the amount lag is so small it is difficult to see comparing signals by eye.

4-4A.   50.
4-4B.   That the procedure finds the correct K_{Est} if the estimated sensor dynamics are just reasonably close to representing the actual sensor.

Chapter 5

5-1A. $K_{DO} = 0.06$, $K_{PO} = 25$, $K_{IO} = 1400$.
5-1B. $K_{DO} = 0.04$, $K_{PO} = 14$, $K_{IO} = 700$.

5-2A. 50.
5-2B. 50.

5-3A. Increased phase lead at and around 50 Hz.
5-3B. Increased phase lag at and around 50 Hz.
5-3C. When *FGs* is lower than nominal, this condition causes phase lag. This can be verified by monitoring the step response shown in the *Live Scope* of Experiment 5D when changing sensor bandwidth.

5-4A. 49° PM and 14 dB GM.

5-4B.

K	Gain C/O (Hz)	PM (deg)	Phase C/O (Hz)	GM (dB)
20	6.4	42	89	30
50	12	49	46	14
100	24	36	43.5	5.8

5-4C.

K	Gain C/O (Hz)	PM (deg)	Phase C/O (Hz)	GM (dB)
20	2.5	41	21	24
50	5	47	21	17
100	9	39	21	11

5-4D. Nominal margins of stability are approximately the same for both the traditional and the observer-based systems. The observer system is more sensitive to increase in plant gain (K) as shown by the loss of 13° PM and 8 dB of GM for K high; compare this to the traditional system, whith a loss of only 6° PM and 6 dB GM. Increased loss of stability in the observer-based system is easily seen in the step response for high K.

Chapter 6

6-1A. 154 Hz.
6-1B. 144 Hz.
6-1C. 103 (for observed disturbance) and 102 Hz.
6-1D. 72 (for observed disturbance) and 71 Hz.
6-1E. They are nearly the same when the estimated plant and sensor are accurate representations of the actual plant and sensor. Under these conditions, Equations 6.4 and 4.6 are identical. *Note: The small difference between 6-1A and 6-1B is caused by variation in the setup of the DSAs (via the "gear" buttons) in Experiments 5A and 6A.*
6-1F. The response of observed disturbance is different from the observer response; Equation 6.4 is not a good approximation when the sensor and plant models vary substantially from the actual sensor and plant.

6-2A.   0.022.

6-2B.   0.05, 0.02, 0.01.

6-2C.   It is inversely proportional.

6-3A.   Should be similar to Figure 6-17 for the second two cases.

6-3B.   4.5, −8.3, and −25 dB.

6-3C.   12.8 dB or 4.4 times.

6-3D.   16.7 dB or 6.8 times.

6-3E.   1.76, 0.404, and 0.06.

6-3F.   1.76/0.404 or 4.35

6-3G.   0.404/0.06 or 6.3.

6-3H.   Both measures produce essentially the same measurements.

6-4A. $K_P=1.5$, $K_I=25$ (50 Hz);
         $K_P=3.0$, $K_I=50$ (100 Hz);
         $K_P=6.0$, $K_I=100$ (200 Hz).

6-4B.   1.2 (50 Hz), 0.6 (100 Hz), and 0.4 (200 Hz).

6-4C.   0.5 (50 Hz), 0.3 (100 Hz), and 0.2 (200 Hz).

6-4D.   No. The majority of disturbance response comes from the decoupling path and the control-law gains have little effect.

Chapter 7

7-1A.   144, 80, 52 Hz.

7-1B.   14 (high), 8 (medium), 0 dB (low).

7-1C.   Yes. K_{DO}.

7-2A.   About 0.2 divisions or 0.04.

7-2B.   About 1 division or 0.2.

7-2C.   About 2 divisions or 0.4.

7-2D.   At 200 Hz, part A (default parameters) −28 dB; part B, −14 dB; part C, −8 dB. Comparison: Part A is 14 dB (5 times) lower than part B, which is 6 dB (2 times) lower than part C. This is consistent with the time-domain measurements.

7-3A.   About 1.2.

7-3B.   About 0.5.

7-3C.   Yes, both in the *Live Scope C* and in the Noise DSA.

7-3D.   8.5 Hz.

7-3E.   Both the noise response and the disturbance response of the systems of parts A and D are equivalent.

Chapter 8

8-1.   First phase crossover disappears when $K_{IO}=0$; without the additional 90° phase lag contributed by K_{IO}, there is no crossover at (or near) 200 Hz.

8-2A. V_{OMod} is highly inaccurate. The other two signals are accurate.

8-2B. V_{OMod} disturbance response will probably be poor because the velocity perturbations caused by the disturbance are so poorly represented by the signal.

8-2C. V_O provides the best response. V_S provides similar response except for strong peaking at 400 Hz caused by the too high control-law gains for the phase lag of the sensed signal. V_{OMod} provides the poorest response as discussed in Part B.

8-3.   The noise sensitivities V_O and V_S are similar. The noise sensitivity V_{OMod} is lower, due to the filtering effect of the modified Luenberger observer (see Section 7.5). *Compare to Exercise 8-2:* The implicit filtering of the modified observer provides advantages (lower noise sensitivity) and disadvantages (poorer disturbance response) equivalent to an observer with a lower bandwidth.

8-4A. 48° in both cases.

8-4B. 24° for V_O and 28° for V_{OMod}.

8-4C. The one using V_{OMod}.

8-4D. 51° in both cases.

8-4E. 40° for V_O and 36° for V_{OMod}.

8-4F. The one using V_O.

8-5A.

T_{SAMPLE}	Maximum K_{AFB}
0.00025	1.2
0.0002	2.5
0.00015	6.0

8-5B.

K_{DD}	Disturbance response at 10 Hz (dB)	Improvement offered by acceleration feedback (dB)
0	−15	*(reference)*
1	−21	6 (2×)
2.5	−26	10 (3.1×)
6	−32	17 (7×)

8-5C.   Ideal result is $1+K_{AFB}$ improvement.
Case 1: $1+1=2$.
Case 2: $1+2.5=3.5$.
Case 3: $1+6=7$.

The results in part B are similar to the ideal results.

Index

Note: Page numbers followed by f or t refer to the figure or table on that page, respectively.

Printed and bound by CPI Group (UK) Ltd, Croydon, CR0 4YY

08/05/2025

01864868-0002